THE OPEN UNIVERSITY

Science: A Third Level Course

**Earth Science Topics and Methods**

Porphyry Copper
Case Study

THE OPEN UNIVERSITY PRESS

# THE S333 COURSE TEAM

**Chairman**

R. C. L. Wilson (*to January* 1976)
D. E. Jackson (*from January* 1976)

**Authors**

S. A. Drury
P. W. Francis
I. G. Gass
D. E. Jackson
L. R. A. Melton
J. A. Pearce
R. S. Thorpe
D. W. Williams
R. C. L. Wilson

**Editor**

F. Aprahamian

**Other Members**

N. Butcher (*Staff Tutor*)
P. Clark (*BBC*)
N. Cleminson (*BBC*)
Beryl Crooks (*IET*)
Patricia McCurry
Valerie S. Russell (*Course Assistant*)
E. Skipsey (*Staff Tutor*)
J. B. Wright

**Consultant**

J. Whitford Stark

The Open University Press
Walton Hall, Milton Keynes
MK7 6AA

First published 1976

Designed by the Media Development Group of the Open University.

Printed in Great Britain by
Staples Printing Group
at St Albans

ISBN 0 335 04333 X

This text forms part of an Open University course. The complete list of items in the course appears at the end of this text.

For general availability of supporting material referred to in this text, please write to the Director of Marketing, The Open University, P.O. Box 81, Walton Hall, Milton Keynes, MK7 6AT.

Further information on Open University courses may be obtained from the Admissions Office, The Open University, P.O. Box 48, Walton Hall, Milton Keynes, MK7 6AB.

1.1

# Contents

**Acknowledgements**

The Course Team are grateful to Dr A. V. Bromley, Dr J. Cole, and Dr W. J. Rea for acting as assessors of earlier drafts of this text. Our gratitude is also due to the many geologists who have assisted us with discussions and comments, in particular, Professor L. B. Gustafson, Dr S. M. F. Sheppard, Mr N. Grant, Dr C. J. Halls, Professor F. J. Vokes and Mr K. R. Greenleaves. Finally, we reserve particular thanks to Dr John P. Hunt, both for his comments on and additions to the text, and for helping *us* to understand the details and implications of the El Salvador study.

Grateful acknowledgement is also made to the following for material used in this *Case Study*:

**Text**

V. D. Perry 'History of the El Salvador development', H. E. Robbins *et al.* 'Development of the El Salvador mine', and W. H. Swayne and F. Trask 'Geology of El Salvador', all of which appeared in *Mining Engineering*, Vol. 12, 1960.

**Figures**

*Figure 1* from R. H. Sillitoe 'A plate tectonic model for the origin of porphyry copper deposits' in *Econ. Geol.*, Vol. 67, 1972; *Figure 2* from *Mineral Resources and the Environment* (COMRATE), reproduced by permission of the National Academy of Sciences, Washington; *Figure 3* reproduced by courtesy of Kennecott Exploration Inc.; *Figure 4* from A. B. Parsons, *The Porphyry Coppers*, AIME, 1933; *Figures 7, 15, 19–22, 26–28, 30, 38–40* from L. B. Gustafson and J. P. Hunt 'The porphyry copper deposit El Salvador, Chile' in *Econ. Geol.*, Vol. 70, 1975; *Figure 10* from D. E. James 'The evolution of the Andes' in *Scientific American*, Vol. 229, 1973; *Figures 11 and 12* from D. E. James 'Plate tectonic model for the evolution of the central Andes' in *Bull. Geol. Soc. Amer.*, Vol. 82, 1971; *Figure 14* from R. H. Sillitoe 'Tectonic segments of the Andes' in *Nature, Lond.*, Vol. 250, 1974, Macmillan and Co.; *Figure 29* based on A. W. Rose 'Zonal relations of wallrock alteration and sulphide distribution at porphyry copper deposits', *Econ. Geol.*, Vol. 65, 1970; *Figure 34* from P. V. Kletsov and G. G. Lemmlein 'Pressure corrections for the homogenization temperatures of aqueous NaCl solutions *Doklady Acad. Sci. USSR*, 1960; *Figure 36* by courtesy of S. M. F. Sheppard; *Figure 37 (a)–(c)* from A. M. Bateman, *Economic Mineral Deposits*, John Wiley; *Figure 37 (d)* T. D. Ford, *Sedimentary Ores: Ancient and Modern* (revised) Proc. 15th Inter-Univ. Geol. Congr. 1967, Dept of Geology, Univ. of Leics.; *Figure 42* from F. H. Howell and J. S. Molloy 'The Braden orebody, Chile', *Econ. Geol.*, Vol. 55, 1960; *Figure 43* from R. H. Sillitoe 'Geology of the Los Pelambres porphyry copper deposit, Chile', *Econ. Geol.*, Vol. 68, 1973; *Figure 44* from R. W. Bamford 'The Mt. Fublian (Ok Tedi) porphyry copper deposit', *Econ. Geol.*, Vol. 67, 1972; *Figure 45* from J. D. Lowell and J. M. Guilbert 'Lateral and vertical alteration – mineralization zoning in porphyry ore deposits', *Econ. Geol.* Vol. 65, 1970; *Figure 46* from R. H. Sillitoe 'The tops and bottoms of porphyry copper deposits', *Econ. Geol.*, Vol. 68, 1973; *Figures 48, 49 and 55* from R. Rice and G. Sharpe *Trans. Inst. Min. Metall.*, Vol. 85, 1976.

**Tables**

*Tables 1, 3 and 6* from *Mineral Resources and the Environment* (COMRATE), reproduced by permission of the National Academy of Sciences, Washington; *Table 5* from J. W. Whitney 'A resource analysis based on porphyry copper deposits', in *Econ. Geol.*, Vol. 70, 1975; *Tables 9 and 10* based on L. B. Gustafson and J. P. Hunt 'The porphyry copper deposits of El Salvador, Chile', in *Econ. Geol.*, Vol. 70, 1975; *Tables 12 and 14* by courtesy of S. M. F. Sheppard.

**Table A1**

**List of scientific terms and concepts taken as prerequisites from earlier Courses**

| | Course/Unit or Block No. | | Course/Unit or Block No. |
|---|---|---|---|
| acid rock | S23–*/1 | metamorphism | S23–/4 |
| aerial photograph | S26–**/3 | metasomatic process | S23–/4 |
| andesite | S100†/26 | mineral assemblage | S23–/4 |
| basic rock | S23–/1 | mineral deposit | S26–/3 |
| batholith | S23–/4 | mineralizing fluid | S26–/3 |
| Benioff Zone | S100/24 | Moho | S100/22 |
| complex ion | S26–/3 | Neptunist | S100/26 |
| concentration of ore | S26–/3 | orogenic belts | S100/24 |
| constructive plate margin | S100/25 | oxidation | S100/8 |
| contact metasomatic deposits | S26–/3 | oxidized zone | S26–/3 |
| | | oxidizing agent | S100/8 |
| continental margin | S100/24 | *P* waves and *S* waves | S100/22 |
| continental rise | S100/24 | permeability | S100/26 |
| continental shelf | S100/24 | pH | S100/9 |
| craton | S100/24 | phase | S23–/4 |
| deductive reasoning | S100/1 | phenocryst | S23–/1 |
| destructive plate margin | S100/25 | plate tectonics | S100/25 |
| disseminated ore | S26–/3 | plate boundary | S100/25 |
| dyke | S23–/2 | Plutonist | S100/26 |
| earthquake zones | S100/22 | porphyritic | S23–/1 |
| erosion | S100/24 | porphyry copper deposit | S26–/3 |
| equilibrium | S100/9 | primary zone | S26–/3 |
| faulting | S100/24 | phenocryst | S23–/1 |
| ferromagnesian minerals | S100/24 | Rb/Sr dating | S2–2††/5 |
| fluid inclusion | S26–/3 | reserve | S26–/1 |
| folding | S100/24 | resource | S26–/1 |
| froth flotation | S26–/3 | secondary enrichment | S26–/3 |
| gangue minerals | S26–/3 | self-potential | S26–/3 |
| gneiss | S23–/1 | seismic waves | S100/22 |
| graben | S23–/2 | sialic crust | S100/22 |
| grade of ore | S26–/3 | soil geochemistry | S26–/3 |
| groundmass | S23–/1 | solid solution | S2–2/4 |
| hydrothermal fluids | S26–/3 | $SO_2$ pollution | S26–/3 |
| inductive reasoning | S100/1 | spectrochemical analysis | S2–2/1 |
| intermediate rocks | S23–/1 | stability field | S2–2/2 |
| joints | S23–/4 | stratiform deposits of Cu–Pb–Zn | S26–/3 |
| K–Ar dating | S23–/3 | stratigraphic trap | S26–/2 |
| Lasky's law | S26–/3 | stream geochemistry | S26–/3 |
| leached zone | S26–/3 | subduction | S100/25 |
| lithosphere | S100/22 | unconformity | S23–/2 |
| magmatic segregation deposit | S26–/3 | vein ore deposit | S26–/3 |
| manganese nodules | S26–/3 | volcanic arc | S100/24 |
| metallogenic province | S26–/3 | | |

*The Open University (1972) S23– *Geology*, The Open University Press (6 Blocks).

**The Open University (1974) S26– *The Earth's Physical Resources*, The Open University Press (6 Blocks).

†The Open University (1971) S100 *Science: A Foundation Course*, The Open University Press (34 Units).

††The Open University (1972) S2–2 *Geochemistry*, The Open University Press (6 Units).

## Table A2

**List of scientific terms, concepts and principles taken as prerequisites from other parts of the S333 Course**

| | S333 Ref*/ Section No. | | S333 Ref*/ Section No. |
|---|---|---|---|
| alkalic rock series | HB/VII | mineral–water isotope fractionation curve | HB/VI |
| conceptual revolutions | MC/4 | oblique aerial photograph | HB/IV |
| delta values | HB/VI | ore texture | HB/VI |
| $\delta D$–$\delta^{18}O$ diagram | HB/VI | oxygen and hydrogen isotopes | HB/VI |
| formation waters | HB/VI | paradigm | MC/4 |
| geothermal waters | HB/VI | polished section | HB/XI |
| geothermometer | HB/VI | primary magmatic water | HB/VI |
| hydrous minerals | HB/VI | reflected-light microscopy | HB/XI |
| hypothetico–deductive reasoning | MC/1 | replacement texture | HB/XI |
| isotopic exchange | HB/VI | sedimentary basin | SB/1 |
| magmatic water | HB/VI | SMOW | HB/VI |
| meteoric water | HB/VI | stable isotope | HB/VI |
| meteoric water line | HB/VI | technological revolution | MC/4 |
| mineral–mineral isotope fractionation curve | HB/VI | tholeiitic rock series | HB/VII |

*The Open University (1976) S333 *Earth Science Topics and Methods*, The Open University Press. The various parts of the Course are referred to as follows: HB, *Techniques Handbook*; MC, *Methods and Consensus in the Earth Sciences*; SB, *Sedimentary Basin Case Study*.

## Table A3

**List of scientific terms, concepts and principles introduced in this Case Study**

| | Page No. | | Page No. |
|---|---|---|---|
| advanced argillic alteration | 44 | microscope freezing stage | 51 |
| argillic alteration | 44 | microscope heating stage | 49 |
| assay | 23 | mineralogical guides to ore | 63 |
| 'blind' deposits | 19 | mineral zoning | 40 |
| block caving | 26 | molasse deposits | 30 |
| borehole logging | 24 | neutralizing power of a rock | 42 |
| breccia pipes | 57 | pebble dyke | 36 |
| calc–alkaline rock series | 30 | physiographic guides to ore | 63 |
| car samples | 26 | primary inclusion | 48 |
| channel samples | 26 | propylitic alteration | 44 |
| chemical controls on ore deposition | 58 | protore | 36 |
| class of copper deposit | 12 | pseudosecondary inclusion | 48 |
| daughter mineral | 48 | red-bed Cu–U–V deposits | 10 |
| $\delta^{18}O$ – temperature diagram | 55 | regional zoning | 31 |
| epithermal deposit | 57 | saddle reefs | 57 |
| exploration borehole | 24 | secondary inclusion | 48 |
| fluidization | 57 | sericitic alteration | 44 |
| form of a deposit | 59 | stockwork | 38 |
| grab samples | 26 | stratigraphic/lithologic guide to ore | 64 |
| homogenization temperature | 50 | structural controls on ore deposition | 57 |
| hydraulic fracturing | 40 | structural guides to ore | 64 |
| hydrothermal deposit | 57 | sulphur isotopes | 56 |
| hypogene | 38 | supergene | 38 |
| igneous breccia | 35 | temperature–fluid composition phase diagrams | 46 |
| *in-situ* leaching | 19 | tin–tungsten deposit | 31 |
| intra-continental basin | 30 | trapping temperature | 50 |
| K-silicate alteration | 44 | underground mapping | 26 |
| massive sulphide deposits | 11 | wall-rock alteration | 42 |
| mesothermal deposit | 57 | | |
| microscope crushing stage | 51 | | |

## Objectives

When you have completed this *Case Study* you should be able to:

1  Define or recognize correct definitions of, and distinguish between true and false statements concerning, the terms, principles and concepts in Table A.

2  By writing a short essay or answering multiple choice questions, describe the history of porphyry copper deposits as a resource and make some predictions regarding their future. (ITQs 3–5, SAQ 6.)

3  By means of a series of short statements explain how advances and revolutions in geological knowledge and technology play a vital role in the exploration and development of new mineral resources. (ITQ 5, SAQs, 7 and 8.)

4  List the important geological features and resource characteristics of porphyry copper deposits. (ITQs 1, 2 and 36, SAQs 1–5.)

5  By writing a short essay or answering multiple choice questions describe the plate tectonic setting, the geological evolution and the distribution of mineral deposits at an Andean type of plate margin. (ITQs 10–12 and 14.)

6  By means of a series of short statements describe the role of the geologist during the exploration, development and production stages of a mine. (ITQs 6–9.)

7  Given a borehole log from a mining prospect, write a short report describing features of economic interest. (ITQ 7.)

8  Given an aerial photograph of part of the Earth's surface, identify possible guides to mineralization. (ITQ 13.)

9  Describe, in a few short statements, the nature and distribution of sulphide minerals in porphyry copper deposits. (ITQs 15 and 17.)

10  Identify hypogene and supergene sulphides found in porphyry copper deposits from photomicrographs of polished sections and make brief statements about their texture. (ITQs 16, 18 and 19.)

11  List the main types of wall-rock alteration, write a sentence describing the characteristic minerals found in each, and identify the type of alteration from photomicrographs of thin sections. (ITQs 20 and 21.)

12  Relate the type of wall-rock alteration found within or around an ore deposit to the possible temperature and chemical composition of the mineralizing fluids. (ITQs 22 and 23.)

13  By writing short statements or answering multiple choice questions, describe and explain the zoning observed in primary mineralization and wall-rock alteration at El Salvador. (ITQs 17, 21 and 24.)

14  Describe in a few short statements experiments which can be used to determine the temperature of trapping, salinity and composition of fluid inclusions, and list the main sources of error and uncertainty in the results obtained. (ITQs 25, 26 and 27.)

15  Given data on oxygen isotopes from coexisting minerals, deduce the temperature of the fluids which were once in equilibrium with these minerals. (ITQ 28.)

16  Given data on hydrogen and oxygen isotopes from hydrous minerals, deduce the isotopic composition of the fluids that were once in equilibrium with these minerals. (ITQ 29.)

17  Given data on hydrogen and oxygen isotopes from hydrous minerals, and information from fluid inclusions, deduce the source of the fluids involved in mineralization. (ITQ 30.)

18  Given a geological map of a mineralized area, point to rock types and structures that could be conducive to ore deposition and give reasons for the choice. (ITQs 32 and 33.)

19  Critically evaluate conflicting models for ore genesis. (ITQ 34.)

20 On the basis of lithological, mineralogical, structural, isotopic and fluid inclusion data for a mineral deposit, construct a hypothesis for its formation. (ITQ 35.)

21 List the geological features of porphyry copper deposits that could be valuable guides in exploration for new deposits and give reasons for the choice. (ITQ 37.)

22 Given information on the geology of a particular area, write a report on the value of carrying out a programme of mineral exploration in that area. (ITQ 37.)

## Study comment

The Sections in this *Case Study* contain study material of varying degrees of difficulty. The table below represents our ideas on the proportion of time that you should spend on each Section. This is, however, only a rough guideline. For example, the *Case Study* builds on *Mineral Deposits*, Block 3 of S26–; if you feel that you need to revise this Block, you may wish to spend more time on Section 2. You should, however, note that Sections 5–8 are the most important and should be allotted the greatest proportion of your time, even if this means omitting a few ITQs in the earlier sections (but if you do this, do read the answers!).

## Relative weights of Sections

(D = difficult, M = moderate, E = easy)

| Section No. | weighting | percentage of time | requires *detailed* knowledge of the following earlier items or Sections |
|---|---|---|---|
| 2 | E to M | 10 | S26–, Block 3; S333, MC |
| 3 | E to M | 5 | – |
| 4 | M | 10 | – |
| 5 | M to D | 30* | Section 4; *Handbook*, Section XI |
| 6 | M to D | 20* | Section 5; *Handbook*, Section VI |
| 7 | M | 10 | Sections 4–6 |
| 8 | M | 15 | Section 2 and 4–7 |

*Includes the study of *Handbook* Sections.

Sections 2 and 6 involve mathematical manipulations and graph plotting. Even if you feel that you are a non-numerate person, you should attempt the exercises in Section 6 as they do lead to a deeper understanding of the origin of the ore deposit. Similarly, you should attempt to understand the chemical equations in Section 5, even if you are not chemistry-orientated. Neither the mathematics nor the chemistry are particularly difficult.

# 1.0 Introduction

Porphyry copper deposits are very large bodies of low-grade copper sulphide ore, usually associated with porphyritic intermediate or acid igneous intrusions. They are economically important because they provide over 50 per cent of the world's copper. They pose some important environmental problems because of the large-scale mining methods that are used to extract the copper. They are also particularly interesting from the geological point of view. In this *Case Study* we shall concentrate on these geological aspects of porphyry copper deposits. In particular, we shall try to find out how these deposits form and also whether geology has a role to play in finding new deposits of this type.

Much of this *Case Study* deals with one particular porphyry copper mine, at El Salvador in Northern Chile. This mine has a special significance. Because of its excellent exposure, it was singled out for detailed geological study by the Anaconda Company, in the hope that this would help in exploring for new deposits. Anaconda equipped the mine with a petrological laboratory and financed a total of 80 man-years of geological mapping. The results of this work were published in *Economic Geology*, by Gustafson and Hunt in 1975, and we have used their data and diagrams as the basis for this *Case Study*.

Section 2 is a general introduction to porphyry copper resources, and this is followed by the El Salvador study. Essentially, the *Case Study* shows how data on one deposit are collected and interpreted. Section 3 deals with the history and development of the mine and shows how mine geologists record their observations. Section 4 covers the geology of the deposit, from its regional setting in the Chilean Andes to the detailed geological maps prepared by the mine geologists. The mineralization is treated in Section 5. We show how the ore minerals are distributed within the deposit, and how mapping and microscope studies can help to unravel a complex history of ore deposition. Also in Section 5 we examine the phenomenon of wall-rock alteration, the mineralogical changes induced in the host rocks by the mineralizing fluids which pass through them. Section 6 is devoted to fluid inclusions and stable isotopes, two types of laboratory study that can enhance our understanding of the ore deposit. The final Sections deal with the implications of the information collected in the previous Sections. Section 7 shows how these data can be synthesized to produce a model for the origin of the ore deposit. Section 8 shows how case studies of this type can provide ideas that can help in the search for new deposits.

This treatment of the El Salvador mine enables us to teach most of the Course Objectives that we outlined in *Methods and Consensus in the Earth Sciences*.

First, we can indicate the various roles of geologists in the investigation and development of mineral deposits. Essentially, the economic geologist has three functions: as an academic geologist, who wants to know why ore deposits occur where they do, and how they are formed; as an exploration geologist, who wants to find the ore; and as a mine geologist, who wants to know how much ore there is, how it is distributed and how it can best be extracted. Their various roles are perhaps best summed up by the story of the academic geologist, the exploration geologist and the mine geologist who together encounter fresh moose tracks in the Canadian bush. 'I'm going to follow the tracks to try to find the moose', said the exploration geologist. 'I'm going to follow the tracks the other way to find out where the moose has come from', replied the academic geologist. While the mine geologist added, 'I'm going to get my gun to shoot the beast.'

Secondly, we can demonstrate the changing rationale behind mineral exploration, from the prospector on horseback who combed the country for signs of ore, to the giant mining companies who employ highly skilled teams of economists, technologists, lawyers and geologists. In the same vein, we can show how advances in mining technology, exploration technique and geological knowledge help us find new deposits to keep pace with a world which is using its resources at an ever-increasing rate.

Thirdly, we can teach some of the methods used to study ore deposits. This includes both 'practical' and 'intellectual' methods. The practical methods include underground mapping, borehole logging, assaying, transmitted-light and reflected-light microscopy, interpretation of phase diagrams, stable isotope studies, and fluid inclusion methods. The 'intellectual' methods include the synthesis of geological, geochemical and structural data on an orebody to produce a model of its genesis, and the ability to predict the location of an ore-body from various regional and local geological 'clues'.

Finally, we should emphasize that the case study approach means that we are using El Salvador only as a method for presenting the general principles involved in the study of ore deposits. The precise details of El Salvador are, therefore, not of prime importance. It is the methods and techniques, the processes, and the principles covered in this *Case Study* that you should try to understand. When you have completed it, you should be able to apply the same approach to any other hydrothermal orebody, including the Cornish tin deposits or the Pennine lead–zinc deposits—two examples that you will come across during this Course.

## 2.0 Porphyry copper resources

**Study comment** This Section introduces porphyry copper deposits by describing their principal geological characteristics, their importance as a resource, their history and their possible future. This serves two purposes: firstly, it provides a means of revising S26–, Block 3 (*Mineral Deposits*) and applying this knowledge specifically to porphyry copper deposits; secondly, it illustrates some of the principles of *Methods and Consensus in the Earth Sciences*, by showing how advances and 'revolutions' in technology, mineral exploration and geology contribute to the search for copper. You should have *Mineral Deposits* available for reference when you study this Section.

### 2.1 The nature of the resource

Most of the basic geological features of porphyry copper deposits have been covered either directly or indirectly in *Mineral Deposits*. To assist you in revising this material and thereby providing the necessary background for tackling this *Case Study*, we have prepared SAQs 1–5 as *revision* SAQs, and you should therefore attempt them before continuing further. Each question has been designed to highlight one particular aspect of porphyry copper orebodies.

SAQ 1 (on types of copper deposit)

Most ore deposits fall into a number of well-defined categories according to the nature of the ore minerals and their mode of occurrence. In Sections 3 and 4 of *Mineral Deposits*, you were introduced to the most important categories, one of which was the porphyry copper deposit (Section 3.5.1). Listed in column A of Table Q are seven different types of copper deposit, all of which are mined somewhere at the present day; you have met the first five of these in *Mineral Deposits*. Match deposits 1–5 with their usual host rock (from column B) and the usual form of the ore (from column C); you can select more than one item from each of the columns. Deposits 6 and 7 are outside your experience but have a go anyway and pay particular attention to the answers.

TABLE Q

| A<br>type of deposit | B<br>usual host rock | C<br>form of the ore |
|---|---|---|
| 1 Porphyry copper (±Mo) | (a) large basic igneous intrusions | (i) large veins |
| 2 Contact metasomatic copper (±Pb, Zn, Fe) | (b) submarine volcanic rocks | (ii) lenses of massive ore |
| 3 Vein copper (±Pb, Zn, etc.) | (c) limestones and dolomites | (iii) disseminated |
| 4 Stratiform copper (±Pb, Zn) | (d) porphyritic intermediate-acid intrusions | (iv) small veins |
| 5 Magmatic segregation copper (±Ni, Fe) | (e) fluviatile sandstones | (v) confined to layers in the host rock |
| 6 Red-bed copper (±U, V) | (f) shallow-water marine sediments | |
| 7 Copper (±Zn, Pb)-bearing volcanogenic massive sulphides | (g) deep-water marine sediments | |

**SAQ 2** (on global setting of mineral deposits)

In Section 5 of *Mineral Deposits*, we explained how certain types of ore deposit are found in particular global settings. Porphyry copper deposits have a distinctive distribution pattern (Fig. 1) occurring in three main belts: a Western Americas belt, an Alpide belt and a SW Pacific belt. Using this diagram, match the porphyry copper deposits (1–6) with the environment in which they were formed (A–D).

1  Sar Chesmah, Iran  *C*

2  Atlas, Philippines  *A*

3  Chuquicamata, Chile  *B*

4  Valley Copper, Canada  *C*

5  Bingham, USA  *C*

6  Bougainville, SW Pacific  *A*

A  a destructive plate boundary of island arc type which is still active

B  a destructive plate boundary of continental margin type which is still active

C  a destructive plate boundary which is no longer active

D  a cratonic area

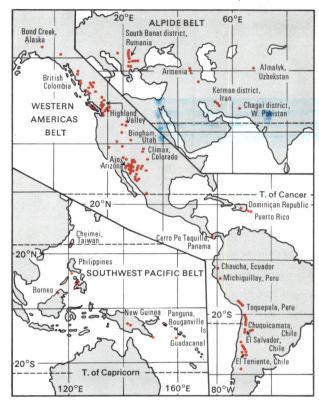

• porphyry molybdenum deposits      • porphyry copper deposits

*Figure 1*  Global distribution of porphyry copper and porphyry molybdenum deposits (from Sillitoe, 1972).

**SAQ 3** (on the origin of porphyry copper deposits)

In Section 3.5 of *Mineral Deposits*, we explained that the sulphide minerals in porphyry copper deposits were deposited from hot, aqueous (i.e. hydrothermal) solutions. For each of the following characteristic features of porphyry copper deposits, indicate whether it provides evidence for this mode of origin?

*No*  1  Pyrite ($FeS_2$), chalcopyrite ($CuFeS_2$), bornite ($Cu_5FeS_4$) and, sometimes, molybdenite ($MoS_2$) are the most common primary minerals.

*No*  2  The ore minerals are distributed fairly uniformly throughout the deposit.

*yes*  3  Most ore and gangue minerals in porphyry copper deposits contain minute fluid inclusions.

*yes*  4  Most ore minerals in porphyry copper deposits are found in a network of small veins, or *stockwork*.

*No*  5  Porphyry copper deposits are genetically associated with igneous intrusions that have a porphyritic texture.

*yes*  6  The mineralization need not be contained within the igneous intrusion but can (and often does) extend into the surrounding country rock.

*yes*  7  The host rocks containing porphyry copper deposits have been extensively altered and often contain abundant secondary hydrous minerals.

**SAQ 4** (on secondary enrichment)

Secondary enrichment has been described only briefly in *Mineral Deposits* (Section 4.5). For most porphyry copper deposits, however, this process is of critical importance. In a number of deposits, only the enriched ore can be profitably mined, whereas the primary, unenriched ore is of too low a grade to be economic. Also, the process of secondary enrichment gives a characteristic colour to the leached rocks at the surface and this can be a valuable aid to exploration. To test your understanding of secondary enrichment, identify whether the following observations apply to the leached zone, the oxidized zone, the enriched zone, or the primary zone of a porphyry copper deposit. Refer to S26–, Block 3, Section 4.5, if necessary.

1 Limonite ($Fe_2O_3.H_2O$) is a common mineral.
*Leached & oxidized*

2 The rock contains chalcopyrite ($CuFeS_2$) and/or bornite ($Cu_5FeS_4$).
*primary*

3 The rock is stained a blue–green colour.
*oxidized*

4 The rock is yellow, red, brown or mixtures of these colours.
*leached*

5 Chalcocite ($Cu_2S$) is the most important Cu-bearing mineral.
*enriched*

6 The rock contains a tiny amount of native copper.
*enriched*

**SAQ 5** (on exploration, mining methods and social implications)

A characteristic feature of porphyry copper deposits is their low grade and large size compared with other types of copper deposit. The average grade of ore being mined at present ranges from 0.5 per cent Cu in Western Canada to about 1 per cent Cu in Chile. Some deposits contain as little as 0.3 per cent Cu. Now decide which of the following features is most likely to apply to (A) a porphyry copper deposit containing 200 million tons of 0.5 per cent Cu ore; (B) a vein copper deposit containing 50 million tons of 2 per cent Cu ore; or (C) equally to both (A) and (B).

*A*  1 The ore must be crushed to a very fine grain size before mineral extraction

*B*  2 Small scale, selective mining methods are more likely to be used

*A*  3 Large amounts of energy must be expended to recover a unit weight of copper

*C*  4 Sulphur dioxide is released as a pollutant during the smelting process

*A*  5 A large capital outlay is needed to get the mine into operation

*B*  6 Gives a steep potential gradient when the self-potential method of exploration is used

By completing these questions you will have built up a picture of some of the features of porphyry copper deposits. However, this picture is still a very simplified one. In the detailed examination of the El Salvador deposit (Sections 3–8), you will gain a much deeper insight into the origin of these deposits and how studies of this type are an important starting point in the search for more ore.

## 2.2   The size of the resource

The importance of porphyry copper deposits as a resource can be gauged from Table 1, which shows that this type of deposit presently contributes over 50 per cent of the world's mined copper. So it is not surprising that many major mining companies are spending a large proportion of their exploration budgets in the search for these deposits. To spend this money most efficiently, a company must have some idea of the size and grade of the deposits it hopes to find.

With this in mind, a number of companies and institutions have carried out a survey of known deposits. The results of one such survey are given in Figure 2. Copper deposits belonging to the three most important categories have been plotted logarithmically on a graph of ore grade against ore tonnage. The diagonal lines on the graph represent a given tonnage of copper metal in a deposit and can be used to divide the deposits into classes according to the amount of copper they contain. The largest classes, 1 and 2, represent the 'super-giant' and 'giant' deposits respectively.

TABLE 1   The contribution, in 1972, of the different types of copper deposit to the world's total mined copper (after COMRATE, 1975).

| Type of copper deposit | Percentage of the world's mined copper |
|---|---|
| Porphyry coppers | 52.4 |
| Strata-bound coppers (in sedimentary rocks) | 26.9 |
| Copper-bearing massive sulphides | 9.9 |
| Others | 10.8 |
| Total | 100.0 |

**class of deposit**

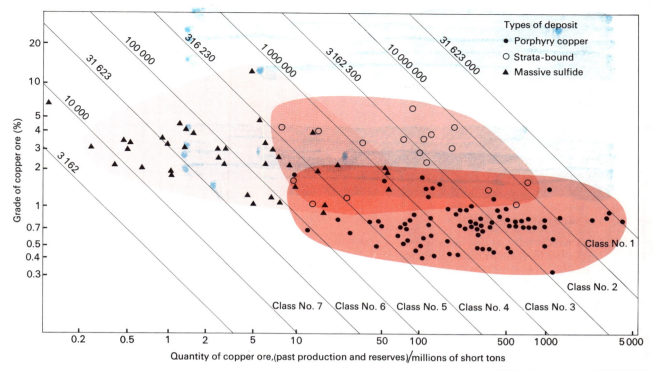

Quantity of copper ore,(past production and reserves)/millions of short tons

*Figure 2* Size and grade characteristics of copper deposits of the three main geological types (see Table 1) (from COMRATE, 1975).

**ITQ 1** Using Figure 2 count the number of porphyry copper deposits in each class. Now calculate the tonnage of copper contributed by each class, by multiplying the number of deposits by the average tonnage of copper in that class. Record your data in Table 2, and use the completed Table to answer the following questions:

(a) Which class of deposit is most commonly found?

(b) Which class of deposit contributes most copper?

(c) Which is likely to be more profitable: a class 1 deposit or a class 2 plus a class 3 deposit?

(d) In what connection have you previously encountered a graph of the type shown in Figure 2?

This exercise has shown that most of the copper lies in a few very large deposits, and it is these that the mining companies are really after. It also emphasizes the huge tonnage and low grades of porphyry copper deposits compared with other types of copper deposit. However, Figure 2 does not indicate the true value of the three types of copper deposit, because it does not include economic factors such as relative costs of extraction, or take into account the value of the by-products which can be extracted along with the copper.

TABLE 2  The contribution to the world's mined copper of the various classes of porphyry copper deposit (to be completed as ITQ 1).

| Class | Average tonnage Cu (approx.) | Number of deposits | Tonnage Cu contributed |
|-------|------------------------------|--------------------|------------------------|
| 1 | $2 \times 10^7$ | 6 | $120 \times 10^6$ |
| 2 | $6 \times 10^6$ | 18 | $104 \times 10^6$ |
| 3 | $2 \times 10^6$ | 30 | $60 \times 10^6$ |
| 4 | $6 \times 10^5$ | 21 | $12.1 \times 10^6$ |
| 5 | $2 \times 10^5$ | 3 | $.6 \times 10^6$ |
| 6 | $6 \times 10^4$ | 1 | $.06 \times 10^6$ |

Gold, silver and molybdenum are the main by-products of porphyry copper deposits. Most porphyry copper deposits contain between 0.01 and 0.05 per cent molybdenum. A few contain as much as 0.1–0.2 per cent Mo and are mined primarily as molybdenum deposits. Gold concentrations vary from less than 0.001 to 0.025 oz/ton (i.e. less than 0.032 to 0.8 parts per million), a substantial quantity when millions of tons of ore are mined each year. Panguna, a porphyry copper deposit on the island of Bougainville in the SW Pacific, makes a substantial portion of its profits from gold, and is, in fact, one of the world's largest *gold* producers. Certain of the rare metals—rhenium (Rh), selenium (Se) and tellurium (Te), which are important as catalysts and are used in alloys for space exploration—are separated from the copper and molybdenum concentrates. Pyrite concentrates (for use in sulphuric acid manufacture) and magnetite concentrates (containing iron ore) are also recovered from some deposits.

### 2.3 The history of the resource

It is a remarkable fact that none of these extremely valuable deposits of porphyry copper had been mined before 1905. The sudden rise in their importance provides a classic example of how technological advances have affected one sub-discipline of the Earth sciences (*Methods and Consensus*, Section 4)·

At the turn of the century, the Bingham mine in Utah was one of many mines extracting small quantities of vein copper ore with a grade of around 6 per cent Cu. Just five years later Bingham had undergone a revolutionary change that was to make it, to this day, one of the largest copper mines in the world (Fig. 3). The mining company's achievement was to apply large scale shovel and railroad mining, as practised in the Lake Superior iron mines, to the exploitation of low grade copper ores (at that time 1–2 per cent Cu). At a stroke, this successful technological adaptation increased the world's reserves of copper by several hundred per cent, for almost overnight it made porphyry copper deposits an economic proposition to mine.

*Figure 3*  The open pit at Bingham, Utah.

The story of the first porphyry copper mine at Bingham makes interesting reading. In his book, *The Porphyry Coppers*, Parsons (1933) described the initial indignities suffered by Thomas Wier, manager of the Boston Consolidated, the founder mining company, who was trying to get stock floated in London. Mr Wier's prospectus computed that within a volume of mineralized ground 2 000 ft long, 5 000 ft wide and 500 ft deep there were 291 666 666 tons of ore, assaying from 0.75 to 2.5 per cent copper. The comments of Richard P. Rothwell, the editor of *Mining and Engineering Journal*, were quoted by Parsons as follows:

'Mr Wier's assumptions are certainly on an exceedingly liberal scale, and it may well be questioned whether a good deal more development work is not needed to prove the existence of so great a mass of ore. But even if we accept the statement and admit that the company has a very large body of low-grade ore, it does not seem to better the case. Even if the greater part—or even the whole—should reach 2 per cent, it would not better the situation. It would be impossible to mine and treat ores carrying 2 per cent or less of copper under existing conditions in Utah.

'In the Montana mines where ores from ½ per cent up are treated, it is well known that the profits come chiefly from the gold and silver in the ores; and it is not claimed that the Boston Consolidated mineral has more than very small values in gold and silver, and many other parallel cases might be presented. Allowing for the greatest cheapness of mining, which can be made possible by the large size of the orebody, one cannot figure out anything but a heavy loss on 1.5 and 2 per cent copper ore. Moreover, it is not probable that the price of copper is going to stay permanently at 17 or 18c. per pound; and with every fall the loss on operations would be greater.

'On the company's own showing, therefore, the more ore it has of the kind it claims, the poorer it is. Undoubtedly, our London friends who are buying the stock at high prices will realize this a little later.'

Parsons then continued:

On one point, the editor (Rothwell) was quite correct; the price of the copper did not stay at 17 or 18c., though the Amalgamated Copper Co., predecessor of the present Anaconda Copper Mining Co., now one of the world's major mining companies, by withholding its output from the market, was successful in sustaining the price during 1901 until November. In that month a precipitous drop sent the price below 12c. per pound.

The point of major significance, however, is that at that period 99 engineers out of 100 who had considered the problem would have frankly held that it was impossible to exploit profitably any deposit containing 40 lb. of copper per ton; and this irrespective of the admitted vast extent of several that were known. The editor unquestionably reflected with accuracy the judgement of the mining engineering profession as a whole.

Perhaps the most tangible factor that they failed to appraise was the economy that might be effected by operation on a large scale. Mass production was not then the potent industrial fetish that it has become. But the fact remains that in 1900 the identical deposits that in recent years have contributed 35 per cent of the world's copper were regarded as so much recalcitrant waste rock!

The entrepreneur and mining engineer who realized the great potential of this deposit was Daniel C. Jackling. By building a concentrator capable of handling huge quantities of ore, and by using steam shovels for stripping overburden and mining the ore, Jackling was soon able to show that the mine could be operated at a substantial profit. Improvements in all aspects of the technology came rapidly, e.g. increase in size of machinery, development of tracked vehicles and, by 1930, electrification. The invention of froth flotation in Britain around this time marked a major breakthrough in concentration of the ore; the operating costs dropped (Fig. 4) and lower grades of ore could be mined.

Bingham's success prompted other mine managers to see whether their mines contained enough low-grade ore to make a similar enterprise worth while. Aided by new techniques, such as rapid drilling methods for helping to work out the

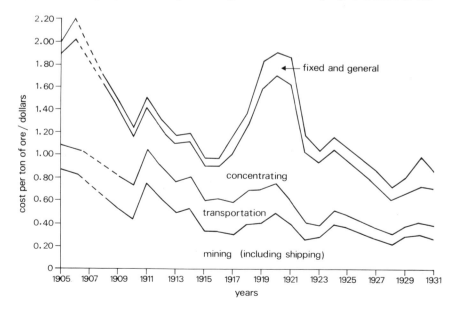

*Figure 4* Production costs at Bingham, Utah. High costs from 1919–1921 are the result of curtailed output during that period (from Parsons, 1935).

size and grade of such orebodies, eighteen more porphyry copper mines were put into production between 1905 and 1956; almost all of these had previously been worked as high-grade deposits on a smaller scale. Figure 5 shows the considerable effect that these porphyry deposits had on world copper production.

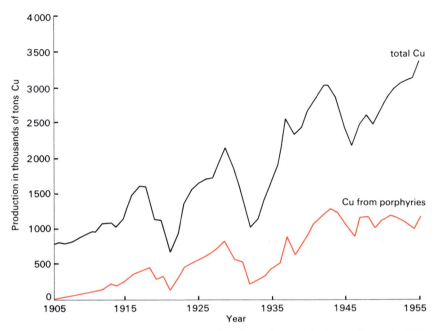

*Figure 5* Contribution of porphyry deposits to world copper production, 1905–1955 (after Parsons, 1935, 1956).

The porphyry coppers brought into production prior to 1956 were from the SW United States (16 deposits) and Chile (3 deposits). Over the next twenty years almost three deposits per year were developed, on average, throughout the world. These discoveries included deposits in completely new areas such as western Canada, the Alpide belt and the SW Pacific. Meanwhile, continuing technological advances such as increases in size of equipment, improved mining efficiency through automation and small improvements in such areas as the design of tyres for mine vehicles enabled lower grades of ore to be mined.

Production figures for the Bingham deposit (Table 3) were compiled by the (US) Committee on Mineral Resources and the Environment (COMRATE) as part of a survey of the world's copper resources. The figures highlight the effects of the technological advances that have taken place since the turn of the century, and also demonstrate the need for frequent updating of reserves and revision of resources forecasts. For example, although in 1914 the size of the Bingham deposit was recorded as $5.5. \times 10^6$ tons of copper; in 1970 the recorded size was $23.4 \times 10^6$ tons of copper.

ITQ 2    From Figure 2, what class of copper deposit is Bingham?

ITQ 3    As we saw earlier, Figure 2 suggests that Lasky's law holds for copper deposits in general. But how valid is it for looking at one particular deposit (*Mineral Deposits*, Section 9.2.4)? Plot the appropriate data from Bingham (Table 3) on to Figure 6, and answer the following questions:

(a) How closely have the copper reserves at Bingham followed Lasky's law during the history of the mine?

(b) Predict the total copper reserves of Bingham mine if it becomes economic to mine an average grade of 0.4 per cent Cu.

(c) Table 3 gives a rate of mining (1970) at Bingham of some 300 000 tons of copper per year. What is the approximate lifetime of the deposit at this rate of mining?

Most known porphyry copper deposits are exposed at the surface. However, many poorly exposed deposits would not have been found but for advances in geological knowledge and new exploration technology. One example of a technological advance was the development in the 1950s of instruments such as the atomic absorption spectrometer for rapid chemical analysis. This enabled the geochemistry of stream sediments and soils to be used as an exploration tool

TABLE 3 Production data for the Bingham mine, Utah, USA (after COMRATE, 1975).

| | Sept., 1899 | Jan. 1, 1915 | Jan. 1, 1930 | Dec. 31, 1970 |
|---|---|---|---|---|
| *Annual production* (prior year) | | | | |
| Tons of waste rock | — | 1 252 961 | 19 821 357 | 97 103 000 |
| Tons of ore mined | — | 6 470 166 | 17 724 100 | 40 147 500 |
| Average Cu grade % | — | 1.4% | 0.99% | 0.68% |
| Tons of copper metal recovered | — | 91 876 | 175 469 | 273 003 |
| *Cumulated production* from start in July 1904 | | | | |
| Tons of waste rock | — | 12 529 611 | 213 195 619 | 1 286 629 883 |
| Tons· of ore mined | — | 34 053 723 | 192 732 074 | 1 170 042 091 |
| Average Cu grade % | — | 1.464% | 1.191% | 0.927% |
| Tons of gross copper contained | — | 498 696 | 2 296 078 | 10 851 559 |
| Tons of copper metal recovered | — | 309 636 | 1 658 464 | 9 745 560 |
| *Reserves—description* | measured ore | ore | ore | proven ore |
| Tons of ore | 12 385 000 | 342 500 000 | 640 000 000 | 1 773 300 000 |
| Average Cu grade % | 2.0% | 1.45% | 1.006% | 0.71% |
| Tons of gross copper contained | 247 700 | 4 966 000 | 6 822 000 | 12 588 000 |
| Tons of estimated copper recoverable | — | — | — | 11 011 000 |
| *Size of deposit* (=past production+reserves) | | | | |
| Tons of ore | 12 385 000 | 376 533 723 | 832 732 074 | 2 943 342 091 |
| Average Cu grade % | 2.0% | 1.452% | 1.094% | 0.796% |
| Tons of gross copper contained | 247 700 | 5 468 696 | 9 118 078 | 23 459 559 |
| Tons of recovered+estimated recoverable copper metal | — | — | — | 20 756 560 |

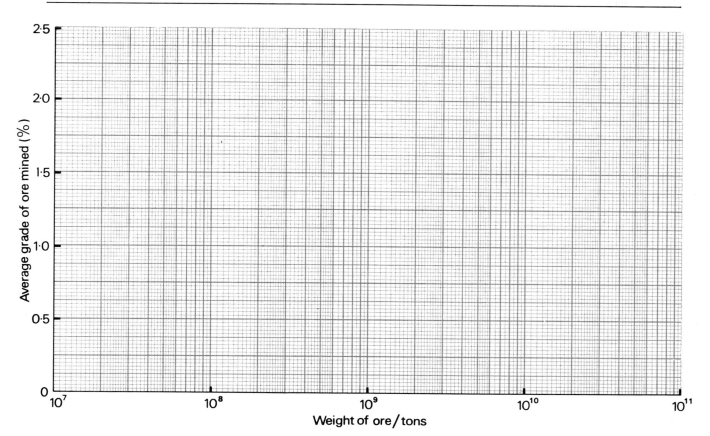

*Figure 6* Graphical representation of Lasky's law for the Bingham deposit (for use with ITQ 3).

(*Mineral Deposits*, Section 6.4.3), and assisted in prospecting for deposits in densely forested areas such as the islands of the SW Pacific. However, neither geophysical nor geochemical exploration techniques have made a very significant contribution to the exploration for porphyry coppers. The state-of-the-art is still largely that some visible evidence of mineralization must be 'sticking out of the ground' in order to reduce the costs of discovery to reasonable levels.

## 2.4 Predictions for the future

The current (1975) rate of consumption of mined copper requires the discovery of one billion ($10^9$) tons of 0.75 per cent copper ore per year, just to maintain world copper reserves at a constant level—equivalent to finding one 'giant' porphyry copper deposit each year (see Fig. 2). So the most relevant question we can ask at the moment is: Where are the copper deposits that have not yet been found or put into production, and which can be developed economically? Our look at the history of the resource may have provided some clues.

Where do you think our copper will come from in the year 2000?

In 2000 AD some copper will undoubtedly come from mines already in production. Many of the existing big copper mines have reserves which will last for 25 years; some have reserves for 50–100 years at least (cf ITQ 3). Some will come from new mines of porphyry copper and other types of copper deposit similar to those mined at the present day, many of which will not be exposed at the Earth's surface or will be found in remote regions. Additional copper will come from deposits that are too low-grade at present to call ore but that will become economically viable as a result of improved technology. Entirely new types of deposit may be mined—manganese nodules (*Mineral Deposits*, Section 4.2.2) are a potentially profitable source of copper, and there may be other sources we cannot yet envisage.

The future of porphyry copper deposits, therefore, depends on two factors: our ability and financial incentive to mine deposits of progressively lower grade; and our ability and incentive to find new deposits. Let us examine these possibilities more closely.

### 2.4.1 Mining lower-grade deposits

The history of the Bingham porphyry copper deposit has shown how improvements in technology can convert apparently low-grade 'waste' rock into copper ore. To what extent this trend will continue is, however, a matter for conjecture; there are limiting factors which may well halt or even reverse the trend, especially in the short term.

Can you think what some of these limiting factors might be?

The first consideration must be whether Lasky's law really does extend to very low grades.

ITQ 4   From your examination of Figure 2 and the results of ITQ 3 do you think it likely that:

(a) there are still significant numbers of 'giant' and 'super-giant' copper deposits to be found?

(b) large porphyry copper deposits presently being worked will prove to contain progressively greater tonnages of copper as mineable grade decreases?

(c) large low-grade deposits other than porphyry coppers will be discovered?

The second limitation lies in the amount of energy required to extract copper from very low-grade ores, a topic that was covered in S26–, Block 6, Section 3.3, to which you should refer for more details.

The total cost in energy terms of mining and processing a tonne of copper ore at 2 per cent grade to the ore concentrate stage amounts to about 120 kWh. That figure includes not only items that most people would think of, like electrical power and fuels, but also the energy required to generate that electricity in the first place, as well as the energy involved in making all the machinery, buildings, and so on, that are involved in the operation. The smelting operation to get copper metal from 1 tonne of this ore concentrate (which contains about 20 kg copper) is also expensive in energy terms, when materials, buildings and so on are included in the sum: about 15 kWh kg$^{-1}$ Cu. So at least 420 kWh of energy must be expended to produce about 20 kg of copper metal from one tonne of 2 per cent copper ore. And the lower the grade, the higher the energy costs will get, although this increase in energy costs is unlikely to be *directly* related to declining grades.

As grades get lower, a higher proportion of the copper is contained in smaller and more finely disseminated mineral particles. How will this affect the energy required to liberate the ore?

The leaner and finer-grained the ore, the more finely crushed the rock must be to achieve good liberation of ore from gangue, so that flotation processes (*Mineral Deposits*, Section 7.4.2) can yield concentrates of reasonable purity. Actually, the crushing energy per tonne of ore is not related linearly to decreasing particle size, but exponentially.

Further limitations to mining lower and lower grade deposits are economic and technological. Both these factors explain how the trend of falling costs of copper production began to reverse after the Second World War (Table 4). Economic factors include increases in the cost of borrowing the large capital sums necessary to finance such a major operation as the development of a large copper mine. At present the main technological factor is the escalating cost, particularly in energy terms, of developing new technology to mine and process lower-grade deposits.

Mining progressively lower-grade ores will therefore demand a new technolological breakthrough with an impact equivalent to that of Jackling's 'revolution' at Bingham and of the invention of froth flotation in England. One such breakthrough may be the successful application of *in situ* or solution mining. *In situ* leaching requires fracturing of the deposit by means of explosives, followed by leaching of copper from the rock by acids assisted by bacterial agents. However, even if this technique proves successful, its world-wide implementation will take both time and capital.

**in situ leaching**

### 2.4.2  The discovery of new deposits

There is, of course, only a finite number of porphyry copper deposits within that part of the Earth's crust which is accessible to us. We do not know how many; the number will depend on factors such as mineable grade, although various estimates have been made using statistical models (e.g. Table 5).

TABLE 5 Simulation data based on historical knowledge showing the possible number of porphyry copper deposits to be found in various regions (from Whitney, 1975).

| | Number of known deposits | Computed most likely number of deposits | Grade range (% Cu) |
|---|---|---|---|
| Canada | 14 | 42 | 0.25–0.95 |
| USA and Mexico | 24 | 106 | 0.3 –1.0 |
| Panama, Columbia and Peru | 8 | 16 | 0.7 –1.0 |
| Chile and Argentina | 9 | 18 | 0.6 –1.8 |
| Philippines | 13 | 74 | 0.25–0.75 |

Exploration for these deposits may be described as involving three general approaches. The most common approach, and that which has dominated the search for ore in the past, is prospecting for deposits which crop out at the Earth's surface. However, this applies only to deposits within that 10 per cent of the surface which is sufficiently well exposed to permit discovery by present prospecting methods. Another common approach involves prospecting around known deposits and within known mining districts. This second approach has been largely responsible for the few discoveries that have been made of so-called 'blind' deposits, i.e. deposits that have no obvious geological, geophysical or geochemical expression at the surface. The most difficult type of exploration involves the search for blind deposits in new areas. To find deposits of this type, which may be covered by a considerable thickness of barren rock, will be one of the great challenges to geologists in future years; it will require the ability to use a knowledge of the geology and origin of ore deposits to predict favourable environments before exploratory drilling and detailed work takes place.

**blind deposits**

TABLE 4 Decline of mineable grade with time and corresponding changes in the cost of mining copper ore.

| Date | Average mineable grade | Cost/lb Cu (in 1970 dollars) |
|---|---|---|
| *1840 BC | 15 | 25 |
| *1540 AD | 8 | 10 |
| 1900 | 4 | 0.64 |
| 1910 | 1.9 | 0.44 |
| 1920 | 1.6 | 0.30 |
| 1930 | 1.4 | 0.33 |
| 1940 | 1.2 | 0.30 |
| 1950 | 0.89 | 0.30 |
| 1960 | 0.73 | 0.37 |
| 1970 | 0.60 | 0.58 |

*Estimates.

There are, of course, limitations on the extent to which we will be able to search for blind deposits. Table 6 shows the trend of exploration costs in the past. The discovery of blind deposits must be accomplished without further great escalations in the cost of mineral exploration. Successful application of geology may well offer the only real possibility of achieving this objective. Massive increases in 'saturation prospecting' (e.g. soil and stream sediment techniques; geophysical testing) have only served greatly to escalate costs without significantly increasing the discovery rate.

It has been suggested that, in order to search for mineral deposits, geologists will need to employ some of the techniques of the petroleum geologist, techniques that you read about in the *Sedimentary Basin Case Study*. Of course, oil deposits make larger targets and are easier to extract. However, as Flawn (1966) says (pp. 364–5):

> The geologist exploring for metals could well study the techniques of the petroleum geologist who has from the beginning been concerned with subjacent (i.e. 'blind') mineral accumulations. In some places he has profited by surface indications such as oil and gas seeps, but for the most part he has sought concealed targets outlined by geologic and geophysical analysis. He has looked not for oil but for geologic conditions which elsewhere have been proved productive of oil and thus his knowledge has grown and been extended on the basis of past success and failure. First he knew only that oil occurred in certain kinds of structures—anticlines and domes. Then he learned about stratigraphic traps and reefs. He found that by building three-dimensional geological models he could find oil, and it is a tribute to his success that he is now probing for blind targets as deep as 25,000 feet below the surface. To build such models he had to understand complex facies relationships and structures within the sedimentary basin being explored, and for this he had to understand the geologic history of the basin.

To apply this kind of approach to the exploration for porphyry copper deposits will certainly require a greater geological knowledge than we already possess, knowledge which might be gained through some sort of geological 'revolution' or through a steady accumulation of geological information (*Methods and Consensus*, Section 4). Perhaps plate tectonic theory will provide the stimulus for such a revolution. You have already seen how plate tectonics has revolutionized other branches of the Earth science, and you will realize from reading S26–, Block 3, that plate tectonics has had some impact on economic geology. But, as every mining company will ask, *will it help to find more ore?*

> ITQ 5 On the basis of your knowledge of the evolution of the plate tectonic theory (from S100 and *Methods and Consensus*) and of mineral deposits (from S26–, Block 3, and this Section), do you think plate tectonics (a) definitely has, (b) probably has, or (c) probably has not:
>
> 1 helped us to understand better the origin of porphyry coppers?
>
> 2 given us a better idea of where to look for porphyry coppers on a regional scale?
>
> 3 given us a better idea of where to look for porphyry coppers on a local scale?

So far, then, plate tectonics has not provided the kind of breakthrough needed to explore for blind porphyry copper deposits. More important may be a steady accumulation of knowledge based on a detailed examination of the characteristics and the local and regional geological settings of known deposits. The Anaconda Company provided an impetus in this direction by financing a detailed geological study of their best-exposed porphyry copper mine, El Salvador in Northern Chile. As mentioned in Section 1, we have chosen this study, published in 1975 by Lewis Gustafson and John Hunt in *Economic Geology*, to demonstrate how a picture of the geology and origin of *one* orebody is built up. We conclude by examining to what extent this and similar studies can help in finding new porphyry copper deposits, and by assessing the limitations of such a predictive model.

> SAQ 6 Assess the following predictions for the future as (a) likely, (b) possible, or (c) unlikely.
>
> 1 The theory of plate tectonics will revolutionize the search for porphyry copper deposits.
>
> 2 The cost of producing copper will continue to increase.

TABLE 6 Changing costs and results of mineral exploration in Western USA (1955–1969) (after COMRATE, 1975).

| Period | Value of metals discovered | |
| --- | --- | --- |
| | Annual average in $ billions | In dollars for each dollar expended |
| 1955–59 | 2.8 | 80 |
| 1960–64 | 3.2 | 59 |
| 1965–69 | 4.1 | 45 |

3 Many porphyry copper deposits will be mined at average grades of 0.2 per cent copper.

4 Future porphyry copper discoveries will contain, on average, less total copper than past discoveries.

5 Porphyry copper deposits in 2000 AD will provide a larger proportion of the total world copper production than at present.

6 The cost of finding new porphyry copper deposits will rise.

**SAQ 7** Is the present state-of-the-art in mining technology

(a) routine, normal science?

(b) normal science, but with a rapid influx of new ideas immediately following a revolution?

(c) a crisis period during which vital new breakthroughs must be developed and applied?

(d) a revolution, with one set of paradigms replacing another?

**SAQ 8** Is the present state-of-the-art in exploration for blind orebodies:

(a) routine, normal science?

(b) normal science, but with a rapid influx of new ideas immediately following a revolution?

(c) a critical period during which vital new breakthroughs must be developed and applied?

(d) a revolution, with one set of paradigms replacing another?

## 3.0  El Salvador: development of the mine

**Study comment**  We begin this in-depth study of one major porphyry copper deposit with an introduction to the history and development of the mine itself. This Section should enable you to understand the role of the mining geologist during the exploration, development and production stages of a mine of this type and also to interpret some of his observations.

The porphyry copper mine at El Salvador is one of the world's largest copper mines. It is situated in the Chilean Andes within a major porphyry copper province and only 20 miles from a similar deposit at Potrerillos (Fig. 7). The commercial orebody at El Salvador constitutes that part of the deposit which has undergone secondary enrichment. It is approximately lens shaped, varying in thickness from a few metres to more than 300 m, and covering an area of approximately 550 m × 1 300 m. The orebody underlies the Indio Muerto Mountain, a prominent peak in the foothills of the Andes, and is mined by an underground bulk-mining method known as block caving. Known reserves prior to production in 1959 were about 300 million tons of ore of average grade 1.6 per cent copper; 80 million tons of this ore were mined during the twelve years of operation under the Anaconda Company before nationalization by the Chilean government in 1971. In this Section, we shall investigate how the mine was discovered, how it was put into operation, and how the vast amount of geological data on El Salvador was collected. A good starting point is to read articles written by the geologists and engineers involved in developing the mine. Extracts from three such articles are reproduced on pp. 22–3.

**ITQ 6**  Read the article by Perry (p. 22), and the extracts from articles by Swayne and Trask (p. 23) and by Robbins *et al.* (p. 23). These articles were all published in *Mining Engineering* in 1960 as part of a section devoted to the mining and engineering aspects of the then newly opened El Salvador mine. While you are reading these articles, list—in the appropriate sections of Table 7—the activities that had to be carried out by mining geologists to get the mine into production. Use the headings given.

*(text continued on p. 24)*

✗ Porphyry copper mine or prospect

*Figure 7*  Location of El Salvador relative to other Chilean porphyry copper deposits (from Gustafson and Hunt, 1975).

# HISTORY OF EL SALVADOR DEVELOPMENT

by V. D. PERRY
Vice President and Chief Geologist,
The Anaconda Co.

The bare hills, the deep canyons and upland slopes of the Chilean Pampa reveal the geology of that region in unique detail. The isolated prominence of Indio Muerto Peak, its significantly dissected sides, its variegated coloring are all surficial features attracting attention to the site of El Salvador mine.

There is no reliable information to explain the few shallow pits and shafts and the turquoise diggings that may go back to early day prospecting in Chile and to early day prospecting in Chile and to Inca or pre-Inca times. Primitive stone hammers and other artifacts, found in a nearby Indian burial ground, testify to the antiquity of some of the workings.

## EARLY OBSERVATIONS

Anaconda's interest, in terms of historical background, is quite recent. It dates from 1922 when, during the early days of development at Potrerillos, I. L. Greninger, then mine superintendent for Andes Copper Mining Co., described a prospect near Indio Muerto Mountain that showed evidence of copper mineralization. Known as the Camp area, it is 1½ m north of Indio Muerto Peak and Turquoise Gulch, under which El Salvador orebody was subsequently found. Greninger's letter stated that the Camp prospect, if located close to transportation, might be worth testing, but that the tonnage possibilities did not appear great enough at that time to warrant construction of transportation facilities.

In the following years of depressed copper prices, there was no interest in the district, and it was not until 1942 that a geological reconnaissance of part of the area was made by Walter March, then Chief Geologist for Chile Copper Co., E. F. Reed, Sam Watson, Jr., and Clifford Wendell, March's assistants. The examination did not fully cover Indio Muerto Peak, and there is no reference in their report to the Turquoise Gulch area. The report recommended further geological examination and the denouncement of claims to cover all mineralized outcrops.

In February 1944, Reno H. Sales, Chief Geologist of Anaconda, March, Wendell and the writer, made a brief visit to the Camp area. Then, in 1945, Anaconda geologists R. B. Mulchay and E. C. Stephens, and Watson of the Andes staff, did further geologic mapping and recommended exploration of the Camp area prospect.

## THE KEY YEAR

Starting about 1950, it became evident that many influential Chileans in and out of government were thinking constructively about long-range and farsighted policies for developing their country's natural resources. Encouraged by this progressive attitude, Anaconda began an intensive exploration effort directed locally to replace the rapidly depleting Potrerillos ore reserves and, generally, to increase Anaconda's participation in Chile's copper production. As part of this program the writer returned to the district in February 1950, accompanied by W. T. Swensen and E. H. Brinley, Andes Company geologists; made a one-day reconnaissance up the northwesterly slopes of Indio Muerto and into the cirque-like amphitheater at the head of Turquoise Gulch. Impressed with the geological structure, rock relations and the character of mineralization and alteration in the leached capping, he made a second trip a few days later, accompanied by William Swayne, Anaconda exploration geologist in Chile.

## THE FIRST PHASE

A letter by the writer, dated April 28, 1950, to C. E. Weed, then Vice President of Anaconda, stressed the theory that the primary mineralizing solutions associated with the Indio Muerto or Turquoise Gulch porphyry carried appreciable amounts of copper within a structural focus of intrusion and fracturing, and that the outcrops of thoroughly leached, barren capping in the vicinity of Turquoise Gulch indicated an underlying zone of secondary chalcocite enrichment. The letter recommended aggressive action in claim denouncement, exploration and development of the prospect. Under the writer's direction, Swayne was assigned the job of surveying and mapping the geology. His enthusiasm and persistence in the face of difficult field conditions were vital factors in constructing an accurate map of the district and in assisting in the selection of initial drilling targets. He was aided in his mapping by John Bain of the Andes' staff and was helped materially by laboratory analyses of systematic rock collections which were studied in the Butte Geological Laboratory by Charles Meyer, then in charge of Anaconda geological research.

## EXPLORATION BEGINS

A detailed report on the district was submitted in May 1951 and, on the basis of the report, plans for exploration were made. Swayne, with the help of an assistant, Hans Langerfeldt, organized, directed, and serviced the drilling job. The first two holes in the vicinity of Turquoise Gulch disclosed thin sections of chalcocite mineralization, and then the drilling site was moved to the Camp area, where several holes developed a relatively small tonnage of primary chalcopyrite ore.

Steady progress was made despite extremely difficult operating conditions, which included rugged terrain, lack of water supply and absence of roads. In April 1954 drills were moved back into Turquoise Gulch, and it was then that important high-grade, secondary chalcocite was cut below the barren oxidized outcrops of the Gulch. Results were so encouraging that it was evident a major discovery had been made.

## THE PROJECT TAKES SHAPE

Events followed in rapid succession, and orders were given by R. S. Newlin, Anaconda Vice President—Mining, to speed up development so that the El Salvador mine could be brought into production at an early date. The program was placed under the direction of C. M. Brinckerhoff, then Vice President of the Chile and Andes companies, and under the immediate supervision of N. F. Koepel, then General Manager, and H. E. Robbins, General Superintendent of the Andes Co. The geological work was kept under the direction of Swayne and the immediate supervision of Frank Trask, Chief Geologist of Andes.

The significant part of the El Salvador story is the continuity of effort on the part of Anaconda geologists, and the coordination of that effort with mine management which, extending over a long period of years, ended in the discovery of an ore body of the first magnitude. Results attained justify the expectation that unremitting efforts on the part of field geologists to record correctly and accurately the factual data of rock relationships, structure, mineralization and alteration, coupled with effective geological research, will produce other major mineral discoveries.

# GEOLOGY OF EL SALVADOR

**WILLIAM H. SWAYNE**
Chief Geologist for
South American Operations

**FRANK TRASK**
Chief Geologist,
Andes Copper Mining Co.

## APPLIED GEOLOGY

... Concentrated geological effort resulted in development of El Salvador orebody. The detailed field work included surface mapping of an area 2 miles wide by 4 miles long. Transit and plane table surveys provided a preliminary triangulation control which was used to establish claim boundaries and check points for subsequent brunton and tape traverses. Field observations of rock types, structures, alteration facies, and leached capping features were recorded in detail. The original field work was supplemented by microscopic studies at the Butte geological laboratory. Coordination between field and laboratory work provided the basis for more detailed field studies upon which a preliminary diamond drilling program was based. Initial holes were inclined from the surface to take maximum advantage of the steep topography. Continued exploration included drilling from tunnels driven in the leached capping and later into the heart of the orebody. Where possible, these holes were directed to penetrate the blanket of secondary ore on 100-meter centers at predetermined horizons. This provided a systematic grid of geologic and assay data upon which the broad plans for mining were laid out, including locations of main entries, ore passes, haulage levels, and ventilation openings.

In conformity with conventional practice all drill cores were split. Half was sent for analysis and the remaining half was subjected to detailed geological mapping and study, in most cases under a binocular microscope. One-half of every piece of core taken at El Salvador is catalogued and filed at the property as a complete record and valuable reference source. Frequent re-studies of these cores have resulted in new ideas beneficial to ore-finding. As of July 1959, 245,113 ft of diamond drill holes were completed with approximately 80 pct core recovery.

All exploration and development headings are mapped geologically and sampled in continuous horizontal cuts. The normal assay inclusion is 5 ft. Car samples of tunnel drives in the ore zone are taken for checks and confirmation of the channel sampling.

All geological information and pertinent sample data obtained from the surface, mine openings, and drillholes, are recorded on both plans and sections. Underground notes are taken on 1:500 scale and recorded on level plans of the same scale. Assay plans are prepared as overlays for the geologic plans. Diamond drill holes are mapped on 1:100 scale and these notes, together with assay data, are recorded on 1:1000 scale cross sections of the individual holes. The sections also show all other penetrations in the area and any significant data pertinent to the layout of ore blocks.

Geological projections and predictions within the mineralized zone, including ore reserve calculations and all ore block planning, are accomplished from 1:2000 scale level plans and coordinated north-south and east-west cross sections of the same scale. These are posted currently with all geological and assay data and serve as basic guides for mining operations.

Geology at El Salvador has been primarily instrumental in the discovery and development of this important porphyry copper orebody. Persistent application of geology will continue to have a fundamentally important part in the problems of efficient ore extraction and in the search for new ore.

# DEVELOPMENT OF EL SALVADOR MINE

**H. E. ROBBINS**
General Superintendent,
Andes Copper Mining Co.

**W. H. DUNSTAN**
Mine Superintendent,
Andes Copper Mining Co.

**T. H. DUDLEY**
Assistant Mine
Superintendent,
Andes Copper Mining Co.

**LLOYD POLLISH**
Mine Foreman,
Andes Copper Mining Co.

El Salvador's rock outcrops are the result of differential erosion which, in general, has left a hard capping of rhyolite. It is expected that this hard capping will inhibit the dilution of ore grade in a caving operation.

Although the enrichment blanket, which constitutes the mineable orebody, is of extraordinary lateral extent and unusual thickness, it is also characterized by local irregularities and by variations in the elevation of the bottom of good ore that are, in places, quite abrupt. Mineable values are overlain by barren material that varies widely in thickness and averages three times the ore thickness.

The following factors influenced the selection of a mining method:

1) An average ore grade of 1.5 pct total copper, with reserves of 375 million tons.

2) Initial production of 12,000 tpd, to be expanded to 24,000 tpd.

3) Ore zones ranging from 80 ft to more than 1600 ft in horizontal dimensions and varying to 740 ft in thickness, with an average thickness of about 425 ft.

4) An irregular ore zone bottom.

5) A column of waste averaging 1300 ft in thickness overlying the ore zones.

6) High degree of fracturing in the mineralized mass.

7) Certain underground mining methods estimated to be cheaper and quicker than open pit mining.

After consideration of these factors, it was decided that block caving with gravity ore disposal was the method best suited to existing conditions. Accordingly, a plan of mining was devised which provided two haulage levels, referred to by their elevations in meters above sea level as the 2600 and 2660 levels. Development was started by driving from surface at both horizons. Each of the two levels, which are operated simultaneously, has two sections separated by an unmineable interval. Each section will eventually have two blocks in full production with other blocks in various stages of development. Under these conditions the mine will have capacity to produce 24,000 tpd on three shifts.

The ore is hauled to either of two ore pass locations for gravity disposal to ore bins feeding a main adit haulage system that leads directly to the coarse crushers at the concentrator.

TABLE 7   The factors required to get a mine into production (to be completed as ITQ 6).

| | |
|---|---|
| (a) Location of the ore | |
| (b) Collection of geological data | |
| (c) Estimation of the grade and extent of the orebody | |
| (d) Choice of mining method | |

You should by now be aware of the enormous amount of effort that must be put in before a mine can produce any ore at all.

This *Case Study* is concerned with the role of the mining geologist but, of course, the economist and the engineer play an equally important part. To put a mine of this size into producton was a major feat of engineering. The engineers at El Salvador were faced not only with the design and construction of the mine, concentrating plant and town, but also with providing adequate supplies of electricity and water to such a remote region. But, to return to the mining geologist, how does he collect the geological data used in this *Case Study*?

His first objective is the construction of a surface geological map. Initially, this will take the form of a reconnaissance map, later a detailed map, of the mineralized area. The methods employed are essentially those used to produce geological survey maps. The scale, however, is by necessity much larger, and accuracy is very important, particularly since these maps will be used to site exploratory boreholes. The key instruments are a plane table, alidade and rod (typical surveyors' instruments), which enable each point in the field to be accurately plotted on to a base map. However, mining geology is primarily a three-dimensional exercise, and logging of boreholes together with underground mapping are a crucial part of the development of a mine. Initially, the objective of borehole logging is to *find* the orebody; later, it is to help in building up a 3–D picture of its grade and extent. Figure 8 shows a borehole log, devised by   **borehole log** the United Nations, for a porphyry copper prospect in Argentina. Major mining companies each develop their own mapping and logging systems and procedures. The data recorded in the UN log include many of the features studied by Anaconda geologists at El Salvador during exploration, development and production of the deposit.

ITQ 7   Using the borehole log in Figure 8:

(a) Mark in column A, the zones of leaching and oxidation, supergene enrichment and primary ore (refer back to S26–, Block 3, Section 4.5, if necessary).

(b) Mark in column B, rough estimates of the average copper and molybdenum grades over the segments 1–4. Write one sentence assessing whether you would recommend further drilling of the deposit on the basis of these estimates: (i) if this were an exploratory hole; (ii) if this were a hole in the centre of the deposit.

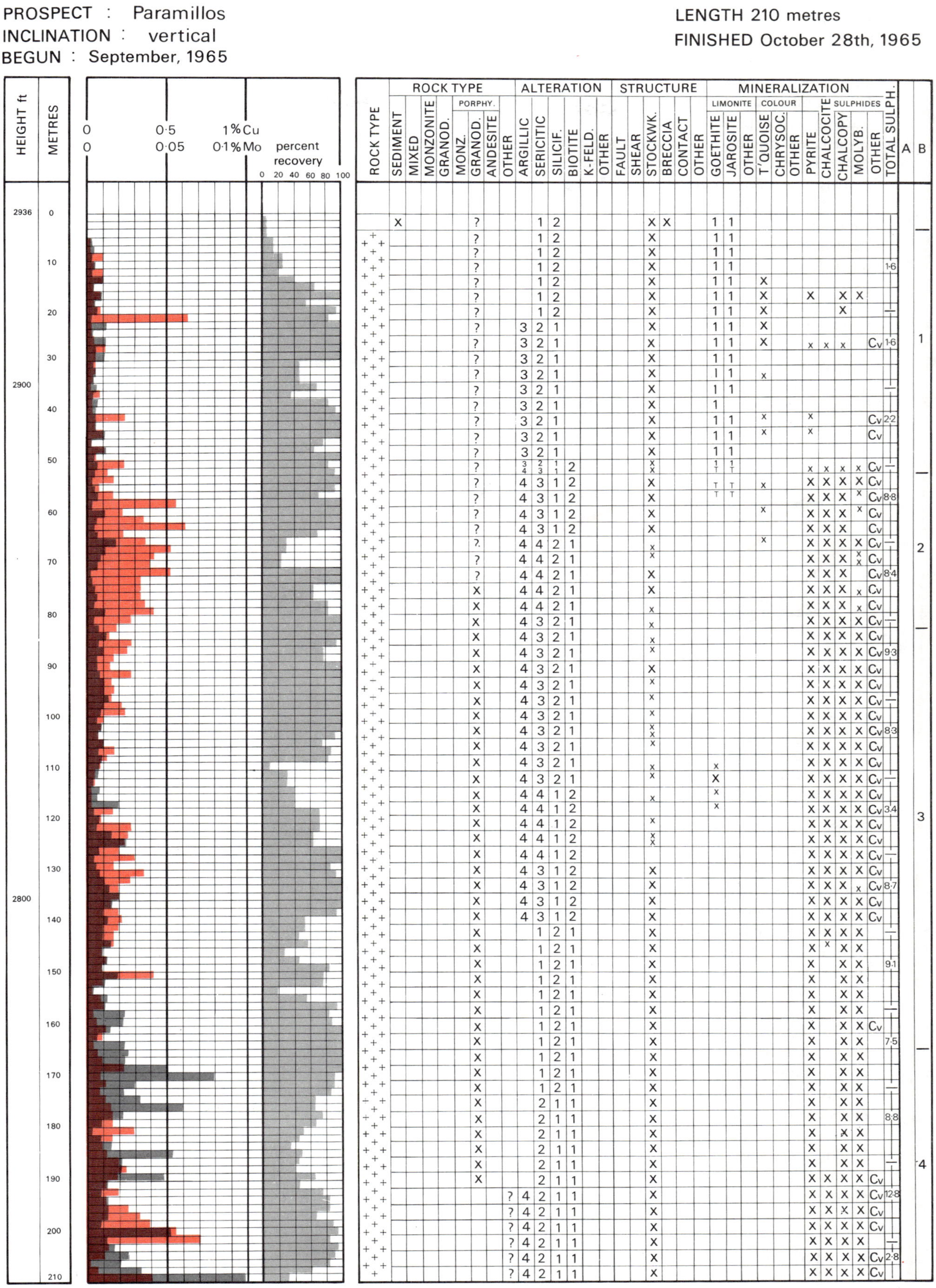

PROSPECT : Paramillos
INCLINATION : vertical
BEGUN : September, 1965

LENGTH 210 metres
FINISHED October 28th, 1965

HEIGHT ft | METRES

0   0·5   1%Cu
0   0·05   0·1%Mo   percent recovery
0  20  40  60  80  100

| ROCK TYPE | ALTERATION | STRUCTURE | MINERALIZATION |

ROCK TYPE — SEDIMENT, MIXED, MONZONITE, GRANOD. | PORPHY.: MONZ., GRANOD., ANDESITE, OTHER

ALTERATION — ARGILLIC, SERICITIC, SILICIF., BIOTITE, K-FELD., OTHER

STRUCTURE — FAULT, SHEAR, STOCKWK., BRECCIA, CONTACT, OTHER

MINERALIZATION — LIMONITE: GOETHITE, JAROSITE, OTHER | COLOUR: T'QUOISE, CHRYSOC., OTHER | PYRITE | CHALCOCITE | SULPHIDES: CHALCOPY., MOLYB., OTHER | TOTAL SULPH. | A | B

Height markers: 2936 (0 m), 2900, 2800, down to 210 m

*Figure 8* Borehole log for a porphyry copper prospect at Paramillos, Argentina (from United Nations). The rock types named are all calc–alkaline igneous rocks; the alteration types (which you will study in detail later in the *Case Study*) are labelled 1–4 in order of intensity. The red and black histograms refer to copper and molybdenum grades respectively.

25

In total, more than 120 km of core from more than 800 diamond drill holes have been logged by the geologists at El Salvador. In addition, probably at least 100 km of underground workings have been mapped in detail. The drill holes and tunnels, together with surface exposure, give a vertical exposure of almost 1 km. Underground mapping is a very precise and systematic exercise requiring training and experience. We cannot teach this sort of skill here, but we can nevertheless attempt to convey some of the principles involved. The geological data is recorded on a base map of a particular level in the mine on which all the tunnels have been accurately located. A very large scale (1 : 500 at El Salvador) is used and, in some vital or very complicated areas, even larger scales (1 : 100) are used. On this scale, individual veins can be mapped. Because mines are often dirty and caked with dust and mud, the walls of the tunnels are sometimes washed and scrubbed before they are mapped so that the geologist can see the rock!

*underground mapping*

Figure 9 is an example of underground mapping in El Salvador, which we include to illustrate the range of information recorded by the mining geologist. Notice the use of a colour code to represent the various minerals and alteration types, and that the relative age relationships and structural features such as faults, dykes and veins are all carefully recorded.

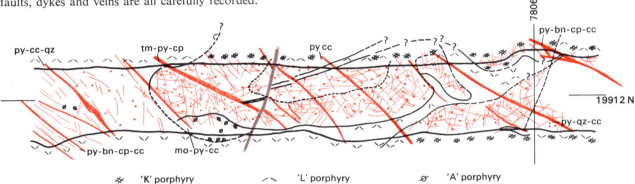

'K' porphyry    'L' porphyry    'A' porphyry

*Figure 9*  An example of geologic mapping of underground workings at El Salvador. The scale is 1 : 100. A colour code is used to represent minerals; here, pink* represents quartz veining, red represents copper sulphides.
*Note that orange, rather than pink, was used in the original map.

**ITQ 8**  The tunnel shown in Figure 9 cuts into two of the main granitic rock types known as the 'L' porphyry and the 'K' porphyry. Contrast these two rock types in terms of (a) intensity of veining (b) age of emplacement and (c) the nature and form of the ore minerals found in each.

*channel samples*

Samples for chemical analysis are taken systematically throughout the deposit from surface rocks, drill cores and underground tunnels. Swayne and Trask (1960) referred briefly to two types of sample: *channel samples*, taken by cutting a continuous series of channels or grooves each 3 m long at chest height on both tunnel walls and collecting the rock chips; *car samples*, collected by taking a shovelful of ore from each car that comes from the orebody. And there are other types of sample that can be taken, such as *grab samples*, from draw points and from the conveyor belts carrying crushed ore in the plant. In fact the assessment of ore grade is based on statistical studies of thousands of analyses.

*car samples*

*grab samples*

In Robbins *et al.* (1960), you read about the block caving method of mining used at El Salvador. The principle of this method is to undercut large blocks of ore, which collapse under their own weight. The resulting crushed rock can be drawn out from below and transported to the metallurgical plant. The mining blocks at El Salvador vary in size from 50 to 100 m in length, and their height is determined by the actual height of the column of ore to be drawn. An important function of the mining geologist is to calculate the tonnage and grade of ore in each block from his geological and assay data. He will also advise the engineers planning the layout of the blocks on other geological features that could influence the mechanical behaviour of the block during mining operations.

*block caving*

**ITQ 9**  What would be your advice to a mining engineer on the expected behaviour during block caving of a planned block of ore, if you found that:

(a) the rock was intensely fractured;

(b) the feldspars in the host rock had been completely altered to clay minerals;

(c) the host rock contained a large proportion of quartz veins;

(d) the host rock was impregnated by anhydrite ($CaSO_4$) causing it to be tough and highly competent?

Write one sentence in each case.

In Section 4, we shall present some of the geological maps compiled from the data of the mining geologists and use these maps to build up a picture of the geological setting of the deposit.

## 4.0  El Salvador: its geological setting

**Study comments**   Detailed studies of the genesis of an orebody can only be carried out when its geological setting has been thoroughly investigated. In this Section, we investigate the setting of the El Salvador deposit on three different scales. On a regional scale, we examine El Salvador in the context of the geology and the geological history of the Andes as a whole. On a local scale, we examine the geological map of the Indio Muerto district in which the El Salvador mine is situated. Finally, we examine the geology of the mine itself, using maps and sections produced by the mining geologists at El Salvador. This Section should provide you with further practice in the interpretation of geological maps and help you to understand some of the processes operating at destructive plate margins, as well as providing the geological background to the later Sections.

### 4.1  The Andes

The Andes have long been one of the world's most important mining areas. They lie along the western margin of the South American plate above the subducted oceanic Nazca plate (Fig. 10) and therefore mark the locus of a destructive plate margin—a region of intense seismic and volcanic activity. In most of Chile, the Andes form a single mountain chain parallel to the coast. North of El Salvador, the mountains branch to form two mountain chains, the Western Cordillera and the Eastern Cordillera, separated by a broad, flat plain, the Altiplano.

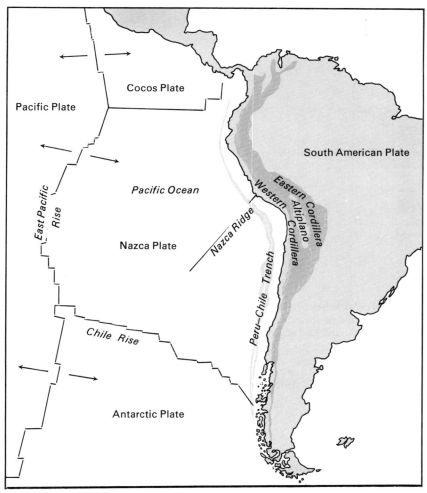

*Figure 10*   The setting of the Andes in a plate tectonic context (after James, 1972).

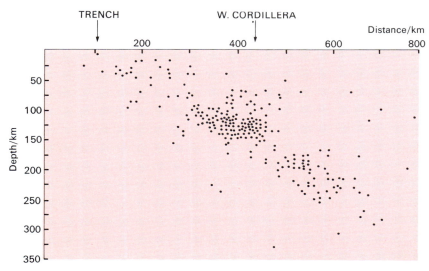

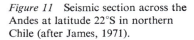

Figure 11 Seismic section across the Andes at latitude 22°S in northern Chile (after James, 1971).

Geophysics has played an important part in our understanding of the structure of the Andes beneath El Salvador. Earthquake foci, located by the World-wide Seismograph Network, have been used to plot a seismic section through the Andes of northern Chile (Fig. 11). The plate boundary lies within the broad band of earthquakes (S100, Unit 22, Section 22.4.6) and dips beneath the Andes at an angle of about 45°. The study of seismic waves also enables us to examine the structure and composition of the Andean crust. Several seismic layers can be recognized (Fig. 12), each characterized by a particular P-wave and S-wave velocity (S100, Unit 22, Sections 22.5.7–8.)

**Andean type of plate boundary**

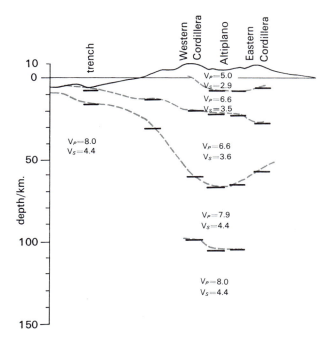

Figure 12 NE—SW cross-section of crust and mantle through the Altiplano in Bolivia. $V_P$ denotes $P$-wave velocity and $V_S$ denotes $S$-wave velocity (after James, 1971).

ITQ 10 Select from items 1–6, the vertical depth beneath the Western Cordillera at which you would be most likely to find the geological features, A–F. Use Figures 8 and 9, together with Table 2 of S100, Unit 22, Section 22.5.11 for this exercise.

| | | | |
|---|---|---|---|
| 1 | 0 km | A | The Moho |
| 2 | 35 km | B | Interbedded sedimentary and volcanic rocks |
| 3 | 65 km | C | A plate boundary |
| 4 | 84 km | D | Metamorphosed oceanic crust |
| 5 | 105 km | E | Partially fused peridotite |
| 6 | 108 km | F | Gneisses |

So, we know that El Salvador is at present situated at an active destructive plate boundary and is underlain by very thick continental crust. However, the deposit itself is of Tertiary age. We, therefore, need to know for how long subduction beneath the Andes has been going on and how this relates to the formation of the El Salvador copper deposit. We must turn to geological evidence for this information.

## 4.2 The geology of northern Chile

In Figure 13, we have reproduced in simplified form the part of the geological map of Chile that relates to El Salvador. You will notice that rocks of all ages from Palaeozoic to Quaternary are exposed. It is now thought that these rocks were formed during two cycles of mountain building: the first during the Palaeozoic, the second from the Jurassic onwards. The Palaeozoic cycle began with continental shelf and continental rise sedimentation in the Ordovician and concluded with a major period of uplift and folding during the Carboniferous and Permian; igneous activity took place throughout this cycle. The second cycle, which is still taking place today, began with submarine volcanism and marine sedimentation during the Jurassic, and evolved to continental sedimentation and igneous activity during and after the Cretaceous. It is only in about the last 10 Ma, however, that the andesitic volcanoes, which now form the crest of the Western Cordillera of the Andes, were built up, and it is also during this time that the Andes were uplifted to their present high altitude. This orogenic cycle appears to have been related to an east-dipping subduction zone which has operated more or less continuously since Jurassic times. We will now examine three important characteristics of this second, economically-important, orogenic cycle: the calc–alkaline igneous rocks; the sedimentation; and the distribution of ore deposits.

**geological history of the Andes**

*Figure 13*  Simplified geological map of the Andes in the region of El Salvador (after the Geological Map of Chile).

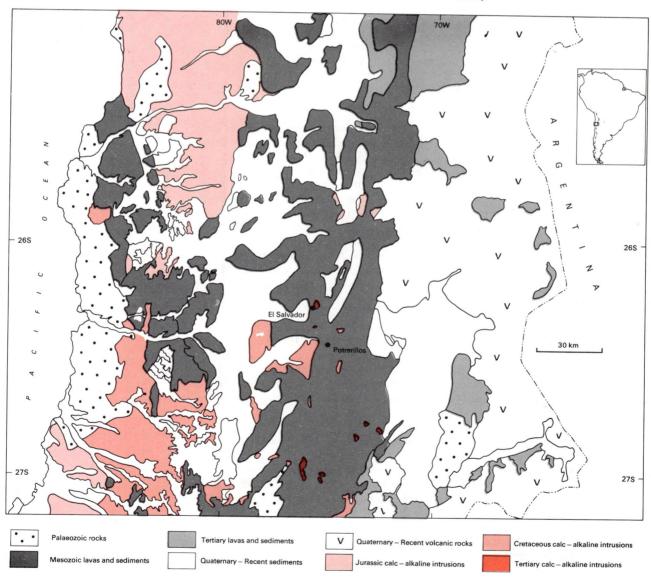

| | | | |
|---|---|---|---|
| · · · Palaeozoic rocks | Tertiary lavas and sediments | V Quaternary – Recent volcanic rocks | Cretaceous calc – alkaline intrusions |
| Mesozoic lavas and sediments | Quaternary – Recent sediments | Jurassic calc – alkaline intrusions | Tertiary calc – alkaline intrusions |

*Calc–alkaline igneous rocks*

The igneous rocks produced at destructive plate boundaries of the Andean type belong to the *calc–alkaline* rock series. They differ in both chemistry and mineralogy from the igneous rocks on Skye, which belong to the tholeiitic and alkalic series (*Handbook*, Section VII). You will be able to identify some of these differences when we look at rocks from the mine later in this Section. Rocks belonging to the calc–alkaline series are classified according to their silica content and to the relative proportions of alkali feldspar and plagioclase feldspar in the rock (Table 8). This classification scheme was used by the United Nations in devising the format for the borehole log in Figure 8.

TABLE 8 Classification scheme for intermediate and acid calc–alkaline volcanic and intrusive rocks.

**Lavas**

| $SiO_2$ | | Alkali feldspar | | Plagioclase feldspar |
|---|---|---|---|---|
| | Quartz+ feldspar | rhyolite | rhyodacite | dacite |
| | Feldspar | trachyte | latite | andesite |

**Intrusive rocks**

| $SiO_2$ | | Alkali feldspar | | Plagioclase feldspar |
|---|---|---|---|---|
| | Quartz+ feldspar | granite | granodiorite | quartz diorite or tonalite |
| | Feldspar | syenite | monzonite | diorite |

Calc–alkaline lavas were erupted in abundance throughout the second orogenic cycle, although their character and position has been continually changing. The lavas have become more acid with time, ranging from Jurassic basalt pillow lavas through to Recent eruptions of dacite and rhyodacite. The locus of the Jurassic volcanic arc lay along the present coastline and must have been partly below sea-level; the Recent volcanic belt is some 150 km inland and is made up of stratovolcanoes up to 6 000 m high. The igneous activity also controls other features of Andean geology such as structures and sedimentation. Intrusion of large batholiths into the continental crust can cause intense folding in the country rocks. The main episodes of folding in the Andes, during the Cretaceous and the Miocene, are also the periods of most intense igneous activity. Sediments in the Andes are mainly derived from the erosion of the igneous rocks, as we discuss further below.

*Sedimentary rocks*

Shallow-water marine sandstones and limestones were deposited during the early stages of the orogenic cycle. Jurassic sediments of this type are found in the El Salvador region. Later sedimentation is continental in character. These sediments, mostly sandstones and conglomerates, were mainly deposited in intra-continental sedimentary basins, often bounded by normal faults and orientated N–S.

What did we call this type of basin in the *Sedimentary Basin Case Study*?

In that *Case Study* (Section 1.1) they were called intra-cratonic graben (Basin type C). In Andean sedimentary basins rock fragments, derived from the erosion of the surrounding igneous terrain, are transported by rivers during the violent floods typical of arid areas and are dumped over the floors of these basins. Continental sediments deposited in this way are termed *molasse* deposits. The largest such basin in the Andes, the Altiplano, was formed during the Cretaceous and now contains a pile of molasse sediments over 15 km thick.

## Distribution of ore deposits

Virtually all the important metallic mineral deposits within the Andes are related to intrusions of intermediate to acid calc–alkaline igneous rocks. Porphyry copper deposits, which occur within and around small intrusions in the Western Cordillera, are the most important economically. Also related to such intrusions, but formed further from them, are vein deposits carrying mainly lead, zinc, copper and silver. Where the intrusions cut carbonate sediments, contact metasomatic iron or copper deposits result. In the Eastern Cordillera, both porphyry and vein deposits are also found, but the main metals are tin, tungsten and silver; Bolivia is one of the world's foremost tin exporters.

> **ITQ 11**  Examine the location of the El Salvador and Potrerillos porphyry copper deposits on the geologic map of northern Chile, Figure 13.
>
> (a) With what age of rocks are these deposits associated?
>
> (b) Why do you think that no similar deposits have been found in the Jurassic rocks along the coastline?
>
> (c) Why do you think that no similar deposits have been found in the Quaternary igneous rocks of the Andes?
>
> (d) One type of deposit which is mined in the Andes but is not obviously related to igneous intrusions is the copper-vanadium red-bed deposit. Where on Figure 13 would you expect to find deposits of this type?

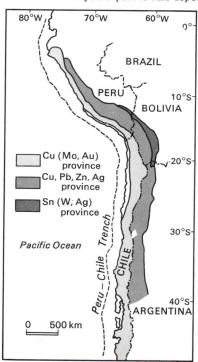

*Figure 14*  Regional zoning of metallogenic provinces in the Andes (after Sillitoe, 1974).

The Andes show a regional zoning of metal provinces (Fig. 14). This results from a variety of factors. As you discovered in ITQ 11, *erosion level* must be sufficiently deep to expose the deposit, but not so deep that the deposit has been eroded away. The *regional distribution of rock types* is also important. As mining geologists say: 'to shoot elephants you go to elephant country'; to find porphyry copper deposits you go to areas containing the appropriate type of calc–alkaline intrusives, not to a sediment-filled intracontinental basin, for example. The *genetic history of the igneous rocks* may also be significant, although we do not yet know in what way.

**regional zoning of ore deposits**

> From your knowledge of the structure of the Andes, name three possible sources for the calc–alkaline igneous rocks.

The igneous rocks could have been produced by the melting of the base of the continental crust *or* the mantle *or* the subducted oceanic lithosphere. Perhaps there are different sources for the igneous rocks related to tin–tungsten deposits and those related to copper deposits. This is, however, only one possible answer to a problem to which we are still seeking a solution.

## 4.3 The geology of the mining district

In Section 4.2 we described some of the main geological features of the Andean plate boundary. In this Section we examine the geological history of one small part of the Andes, the Indio Muerto district, and then look more closely at the geology of the mine itself.

### 4.3.1 The geology of the Indio Muerto district

Look first at Figure 15, a surface geological map of the Indio Muerto district in northern Chile. The El Salvador mine is situated within the igneous intrusive rocks of the 11 000 ft Indio Muerto Peak in the centre of the map. The surrounding rocks are andesitic volcanics and continental sediments. The main Andean volcanic chain lies some 100 km to the east of this area.

> **ITQ 12** Using Figure 15, determine, where possible, the period(s) during which the following features were formed:
>
> (a) the Hornitos unconformity
>
> (b) the Indio Muerto unconformity
>
> (c) a major period of faulting
>
> (d) N–S trending faults
>
> (e) intrusion of igneous rocks
>
> (f) volcanic activity

Your answers to ITQ 12 should have helped you to build up a picture of the Indio Muerto district which can be summarized as follows:

1  Eruption of andesites and deposition of continental sandstones and conglomerates during the Upper Cretaceous.

2  Folding and faulting about a northerly-trending axis, probably during the Lower Tertiary.

3  Eruption of at least two series of andesite and rhyolite lavas during the Tertiary period.

4  Intrusion of a number of intermediate to acid igneous stocks and dykes during part of the Tertiary (from 50 to 41 Ma).

5  Erosion to the present landscape with deposition of gravels of Miocene and Pliocene age.

In an area like the Andes, where exposure is excellent, aerial photographs are commonly used, both for mapping and for the identification of features which might indicate mineralization. Two such features can be seen in Figure 16, a high oblique aerial photograph of the Indio Muerto Peak. They are the reddish-brown leached capping which forms above secondarily enriched ore, and the bleaching caused by the alteration of country rocks by the mineralizing fluids. In Figure 17 we have 'cut open' the mountain to show where the actual orebody is located.

> **ITQ 13** By comparing the aerial photograph (Fig. 16) with the geological map (Fig. 15) of Indio Muerto, locate on the photograph the following features:
>
> (a) Indio Muerto peak
>
> (b) Cerro Pelado
>
> (c) Cretaceous andesites
>
> (d) rhyolite hill
>
> (e) Hornitos unconformity
>
> (f) leached capping
>
> (g) outer limit of visible alteration
>
> *Remember when you attempt this exercise that the aerial photograph was taken facing south-eastwards.*

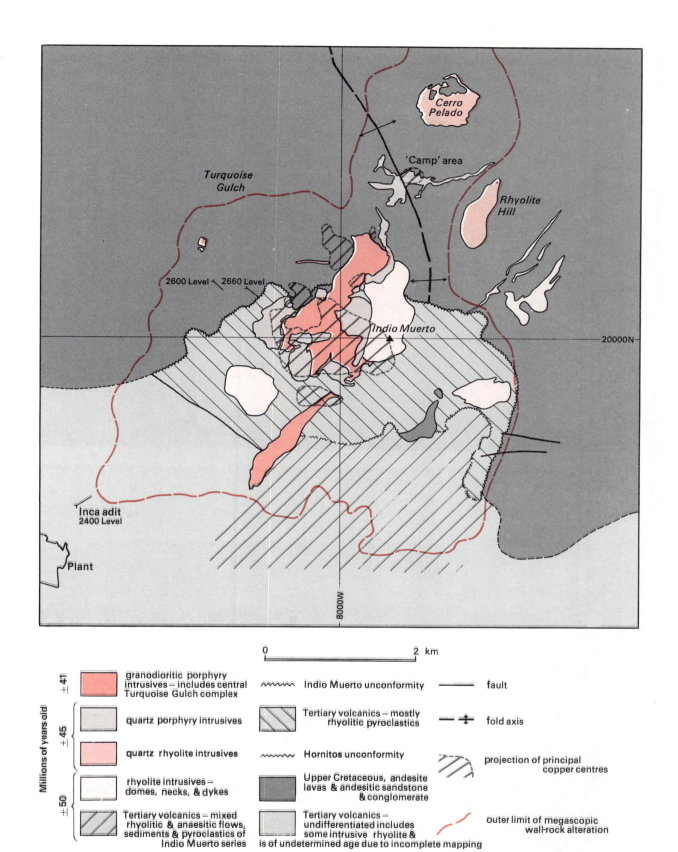

Figure 15  Simplified geological map
of the Indio Muerto district, Chile
(from Gustafson and Hunt, 1975).

33

*Figure 16* Aerial view of the Indio Muerto peak looking south-east directly up Turquoise Gulch. The top of the mountain is a reddish-brown; the rocks in the foreground are green–brown.

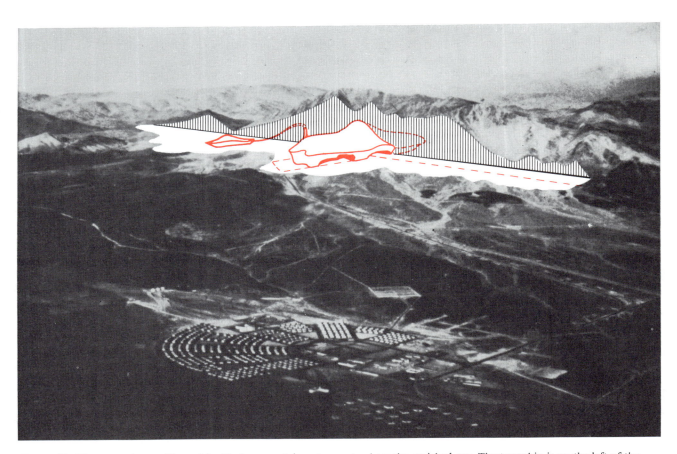

*Figure 17* The same view as Figure 16 with the mountain cut away to show the enriched ore. The township is on the left of the photograph, the concentrator on the far right.

### 4.3.2 The geology of the mine

The Anaconda geologists have built up a detailed three-dimensional picture of Turquoise Gulch, the part of the Indio Muerto district in which the deposit is situated. Their main problem involved the identification of the different types of intrusive rock and the establishment of their age relationships. By careful mapping they recognized three main calc–alkaline rock types.

**1 Indio Muerto, rhyolite domes and dykes**
These are light-coloured, flinty and locally-banded rocks with phenocrysts of alkali feldspar.

**2 Quartz porphyry stocks and dykes**
These contain quartz and plagioclase feldspar phenocrysts in a fine-grained groundmass.

**3 Granodiorite porphyry stocks and dykes**
These are the most important rock types because they are most closely related to the mineralization. Gustafson and Hunt were able to examine contacts within the granodiorite porphyries and subdivide them into a number of distinct types, named 'X', 'K' and 'L' after the cross-cuts in the original exploration tunnels where they were well exposed. The *X*-porphyry was described as a fine-grained granodiorite commonly with an equigranular texture; the *K*- and *L*-porphyries are characterized by phenocrysts of plagioclase feldspar, mafic minerals (biotite and hornblende) and, rarely, quartz.

> **ITQ 14** Table 9 gives some typical analyses of the *L*-porphyry from El Salvador.
>
> (a) Plot these analyses on to the AFM (alkali, iron, magnesium) diagram (Fig. 18). Which shows the greater iron enrichment, the El Salvador porphyries or the Syke tholeiitic lavas?
>
> (b) Do the El Salvador porphyries or the Skye epigranites (*Igneous Case Study*, Table 5) have (i) a greater volume, (ii) more $SiO_2$, (iii) a greater proportion of mafic minerals?

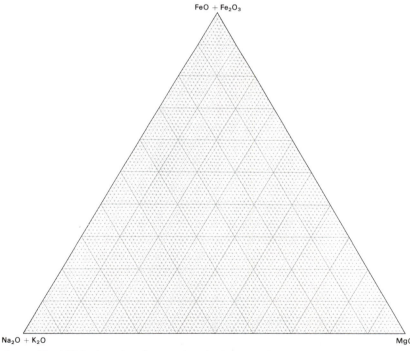

*Figure 18*  AFM triangular diagram for plotting the El Salvador igneous rocks (for use with ITQ 13).

TABLE 9  Three typical analyses of fresh to slightly altered *L*-porphyries from El Salvador (from Gustafson and Hunt, 1975).

|  | 1 | 2 | 3 |
|---|---|---|---|
| $SiO_2$ | 56.58 | 64.31 | 65.09 |
| $Al_2O_3$ | 17.41 | 16.29 | 15.03 |
| $Fe_2O_3$ | 3.44 | 2.63 | 2.05 |
| $FeO$ | 2.72 | 1.77 | 1.27 |
| $MnO$ | 0.03 | 0.03 | 0.02 |
| $MgO$ | 2.15 | 1.60 | 1.31 |
| $CaO$ | 6.14 | 4.34 | 3.87 |
| $Na_2O$ | 4.65 | 4.79 | 3.56 |
| $K_2O$ | 1.57 | 1.79 | 2.68 |
| $TiO_2$ | 0.73 | 0.71 | 0.48 |
| $P_2O_5$ | 0.48 | 0.32 | 0.22 |
| $H_2O^+$ | 1.18 | 0.98 | 2.47 |
| $CO_2$ | 0.33 | 0.24 | 0.35 |
| Total | 97.41 | 99.80 | 98.41 |
| $Na_2O+K_2O$ | 43.8% | 53.4% | 59.0% |
| total iron (as FeO) | 41.0% | 33.6% | 28.6% |
| $MgO$ | 15.2% | 13.0% | 12.4% |

In addition to the three granodiorite porphyries (*X*-, *K*- and *L*-porphyries), there are several minor intrusions which help us to understand the evolution of the ore deposit. Two such intrusions are the *igneous breccias* and *pebble dykes*, typical exposures of which can be seen in Figure 19. The igneous breccias are made up of heterogeneous angular rock fragments in a fine-grained igneous matrix. They probably formed during the final surges of magma following emplacement of the last major intrusive. The pebble dykes have rounded fragments within a clastic matrix. They result from a fluidization process, which

**igneous breccias**

**pebble dykes**

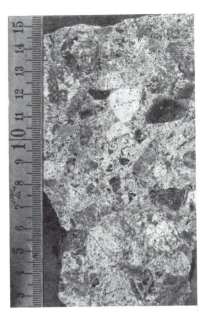

*Figure 19* Pebble dykes. (A) Surface exposure showing rounded pebbles in a sandy clastic matrix. (B) Sawed specimen of a pebble dyke from the deepest level (from Gustafson and Hunt, 1975).

causes abrasion and rounding of rock fragments and is driven by rising steam formed by injection of the last phases of magma into ground-water charged rocks. Dykes of *latite* (see Table 8), a third minor intrusive, make up this final event.

Two of Gustafson and Hunt's maps of the mine have been reproduced in Figures 20 and 21. Figure 20 shows a geological map of the 2 600 m level in the mine (i.e. 2 600 m above sea-level); maps of this type were prepared for all the

*Figure 20* Rock types at El Salvador – 2 600-metre mine level (from Gustafson and Hunt, 1975).

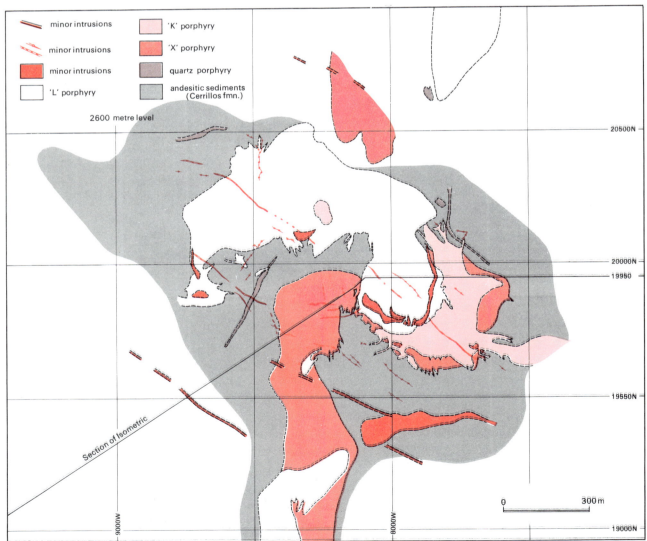

mine levels. Figure 21 is an isometric section through the mine compiled from the surface and underground geological data.

*Carefully study Figures 20 and 21. By examining the contacts between the intrusive rocks listed in the key to Figure 21, attempt to deduce their order of intrusion.*

Radiometric dating (Table 10) helped to quantify the relative age relationships deduced from the mapping, although the various granodiorite porphyries proved to have too similar ages to be distinguished in this way. Gustafson and Hunt were able to discover that the igneous activity at El Salvador covered a 9 Ma period between 50 and 41 Ma. The mineralized intrusives were intruded at the end of this period, first the *X*-porphyry, then the *K*-porphyry and finally the *L*-porphyry.

TABLE 10   A summary of the radiometric ages for igneous rocks from El Salvador (after Gustafson and Hunt, 1975).

| Rock type | Method | Age (ma BP) |
|-----------|--------|-------------|
| Indio Muerto series volcanics | Rb–Sr | $50.3 \pm 3.2$ |
| Indio Muerto rhyolite domes | Rb–Sr | $50.4 \pm 2.8$ |
| Quartz rhyolites | Rb–Sr | $45.4 \pm 1.4$ |
| Granodiorite porphyries | Rb–Sr | $41.4 \pm 1.1$ |
| | K–Ar* | $41.4 \pm 2.6$ |
| Latite | K–Ar** | $40.9 \pm 0.5$ |

\* Dating carried out on hornblende, biotite and sericite separates.
\*\* Dating carried out on biotite separates.

Figure 21   Rock types at El Salvador seen in isometric projection (from Gustafson and Hunt, 1975).

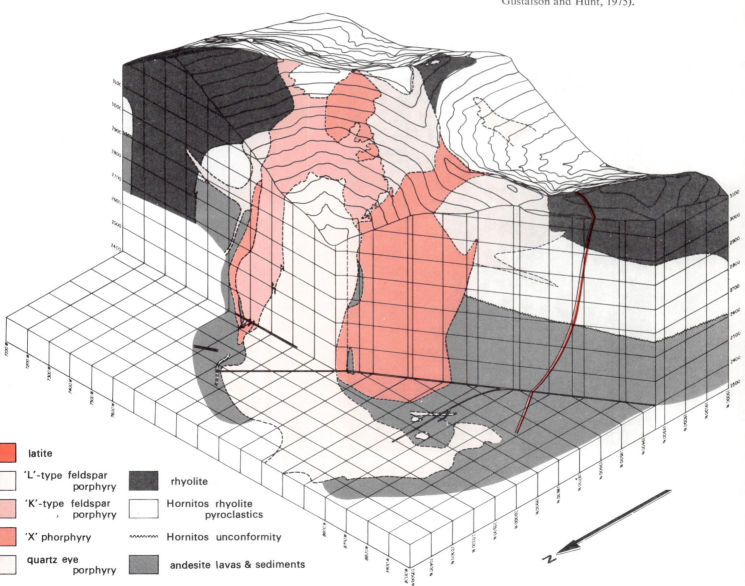

latite

'L'-type feldspar porphyry          rhyolite

'K'-type feldspar porphyry          Hornitos rhyolite pyroclastics

'X' phorphyry          ∿∿∿ Hornitos unconformity

quartz eye porphyry          andesite lavas & sediments

## 5.0   El Salvador: mineralization and alteration

**Study comment**   In the previous Section you were introduced to the geology and rock types at El Salvador, and you learned that the ore minerals were related to a series of intrusions of porphyritic granodiorite. Now, we are going to look more closely at these mineralized rocks, firstly at the ore minerals they contain (*the mineralization*) and then at the minerals produced by reaction between ore-bearing fluids and the fresh host rocks (*the wall-rock alteration*). You may find this Section quite difficult because it contains a number of mineral names and geological terms that are new to you, and because El Salvador—like most ore deposits—has had a complex history that has to be unravelled. You should therefore not worry if you do not assimilate all the *details* of the El Salvador mineralization, but should try to gain an understanding of the *processes* that take place when hydrothermal fluids circulate through rock to form an ore deposit.

### 5.1   Primary mineralization

You will remember from Section 3 that only the secondary enrichment zone can be economically mined at El Salvador at present. However, it is the primary, unenriched ore minerals that are of most interest in interpreting the origin of the deposit. These minerals, produced during the original period of mineralization are termed *hypogene*, whereas the minerals which were formed by later secondary enrichment are termed *supergene*. The volume of presently uneconomic hypogene sulphides is termed *protore;* it may, of course, become economic in the future.

**hypogene and supergene minerals**

**protore**

*Figure 22*   A typical veined rock from El Salvador showing a continuous vertical vein (*B*-vein) cutting less continuous lacing veins (*A*-veins) which carry abundant disseminated sulphides (after Gustafson and Hunt, 1975).

The primary mineralization at El Salvador took place over a time-span of less than one million years and was related to the intrusion of the porphyritic (*X*, *K* and *L*) granodiorites. The mineralizing fluids gained access to these and the surrounding rocks by means of numerous fractures, and the resulting texture characterized by countless criss-crossing veinlets is known as a *stockwork*. The fracturing may be related to the crystallization and cooling of the igneous intrusion, or it may be caused by explosive shattering during the expulsion of fluids from the cooling magma.Fracture densities of tens to hundreds of fractures per square metre are observed in porphyry copper deposits (Fig. 22).

**stockwork**

There are two types of mineralization at El Salvador: *vein* mineralization, in which the sulphides are deposited along with quartz and other gangue minerals in continuous fractures; and *disseminated* or background mineralization, in which the sulphides are deposited in the matrix of the country rock or in small discontinuous fractures. These have different characteristics and we examine them separately below.

### 5.1.1 Vein mineralization

You have already seen in Figure 22 just how abundant sulphide-bearing quartz veins are at El Salvador. Detailed mapping of these veins and their cross-cutting relationships enabled the mine geologists to recognize three distinct classes of vein, which they named 'A' veins, 'B' veins and 'D' veins (initially 'C' veins were also mapped, but these were later found to belong to the other classes). The *A*-veins are the oldest and mark early mineralization at El Salvador; *B*-veins mark transitional mineralization; and *D*-veins, the youngest, mark late mineralization.

ITQ 15   Figure 23 shows a typical veined rock from El Salvador. Which of the two quartz veins is the younger, vein V or vein W?

*Figure 23*   Quartz veins in an andesite from El Salvador (for use with ITQ 15).

The three types differ in structural style, in the minerals they contain and in the alteration of wall rock that has taken place around them. Let us briefly examine some of these differences.

The *A*-veins were formed throughout the period of porphyritic granodiorite intrusion, from before the emplacement of the *X*-porphyry until after emplacement of the *L*-porphyry. The *A*-veins are mostly discontinuous and segmented, with irregular walls (Fig. 24) although the youngest *A*-veins do occupy more continuous fractures. The gangue minerals in the veins are quartz, alkali feldspar, biotite and anhydrite ($CaSO_4$). The sulphide minerals are chalcopyrite and bornite.

By contrast, the *B*-veins postdate the period of porphyritic granodiorite intrusion. Typically they are continuous, nearly horizontal veins with parallel walls and some internal banding. Quartz is the predominant gangue mineral and molybdenite the most common sulphide (Fig. 22).

*D*-veins postdate all rock types except the latite dykes. They are of variable thickness and occupy continuous fractures that are oriented in a definite radial pattern within the deposit. Quartz and anhydrite are the main gangue minerals. Pyrite is by far the most abundant sulphide mineral, but some chalcopyrite, bornite, enargite, tennantite, sphalerite and galena have also been identified. The *D*-veins are sourrounded by a halo or envelope of altered rock, where the fluids that migrated up the fracture caused mineralogical changes in the surrounding rock.

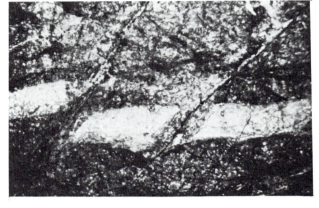

Figure 25   A dyke of *L*-porphyry.

Figure 24   Interlacing *A*-veins at El Salvador.

The mineral, sericite, which you will meet in Section 5.3, is an important constituent of these halos. The *D*-veins are thus the only ones to have significant *wall-rock alteration* (Fig. 25).

This detailed analysis of the vein system that was made by Gustafson and Hunt in their paper emphasizes several important points. It shows that the formation of the deposit was not a simple one-stage process, that the chemistry of the mineralizing fluids changed with time, and that these fluids caused changes in the mineralogy of the rocks they passed through. It also reveals that the fracture system controlling the flow of the hydrothermal fluids changed in character during the evolution of the deposit. Probably the irregular fractures forming the *A*-veins resulted from hydraulic fracturing of the granitic rocks as fluid pressure built up within the crystallizing intrusive. The *B*-vein and *D*-vein fractures developed later, as a response to cooling or release of pressure due to erosion.

**hydraulic fracturing**

### 5.1.2   Disseminated mineralization

Disseminated sulphides—those not found in the vein system—contain between two-thirds and three-quarters of the total copper in the El Salvador deposit. An important feature of the disseminated mineralization is the *zoning*, the spatial distribution of sulphide minerals within the deposit. Zoning reflects the changing character of the mineralizing fluids during the evolution of the deposit. To study the zoning, the field evidence is supplemented by examining the sulphides in polished section under the microscope. Quantitative estimates of sulphide mineral proportions are obtained by systematic point counts (*Handbook*, Section VII) of many polished sections.

**zoning of minerals**

*Now read the Handbook, Section XI, on reflected-light microscopy.*

Now you should be able to identify some of the ore minerals found at El Salvador and interpret some of the textures, using the photomicrographs of polished sections (Slides 1–6) provided with the Course. Let us start with identification of the minerals and examination of the zoning.

> **ITQ 16**   Slides 1–3 are photomicrographs of polished sections taken under reflected light with the analyser out. The minerals present are pyrite, chalcopyrite and bornite.
>
> (a) For each slide, identify which of these minerals are present, using Table 1 of the *Handbook*, Section XI.
>
> (b) These three slides are representative of disseminated mineralization at the 2 600 m level in El Salvador mine. Slide 1 is representative of mineralization near the centre of the deposit; Slide 2 of mineralization further out from the centre; and Slide 3 of mineralization at the periphery of the deposit. What conclusions can you reach about the zonal distribution of Cu : Fe and the metal : sulphur ratios in sulphide minerals within the deposit?

Gustafson and Hunt summarized the results of their point counts by plotting the primary disseminated mineral assemblages on a cross-section through the mine and finally on an isometric section, the same section as that used to show the rock types in Figure 1. They were able to do this even for higher levels, where the sulphides had largely been destroyed by leaching and secondary

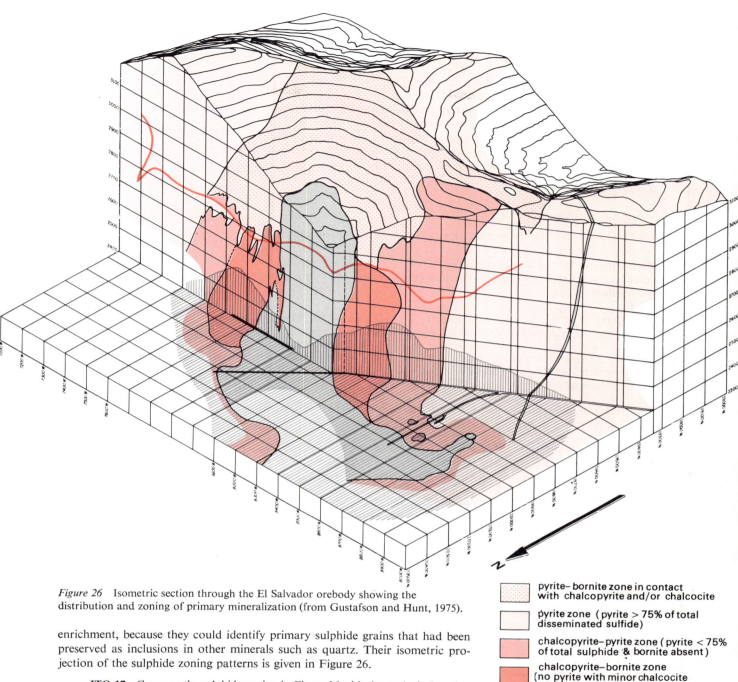

*Figure 26* Isometric section through the El Salvador orebody showing the distribution and zoning of primary mineralization (from Gustafson and Hunt, 1975).

**Legend:**

- pyrite–bornite zone in contact with chalcopyrite and/or chalcocite
- pyrite zone (pyrite > 75% of total disseminated sulfide)
- chalcopyrite–pyrite zone (pyrite < 75% of total sulphide & bornite absent)
- chalcopyrite–bornite zone (no pyrite with minor chalcocite in high-bornite central zone)
- low-sulphide zone
- top of sulphide
- sulphate zone

enrichment, because they could identify primary sulphide grains that had been preserved as inclusions in other minerals such as quartz. Their isometric projection of the sulphide zoning patterns is given in Figure 26.

**ITQ 17** Compare the sulphide zoning in Figure 26 with the geological section in Figure 21.

(a) In what rocks is the mineralization found?

(b) How do you explain the absence of mineralization at the core of the deposit (the low-sulphide zone in Fig. 26)?

Figure 26 shows that there are four main sulphide zones: a central zone of *chalcopyrite and bornite*; a zone of *chalcopyrite and pyrite*; a peripheral zone of *pyrite*; and an upper zone of *pyrite and bornite*. There is also a zone of sulphate mineralization, where anhydrite ($CaSO_4$) is an important gangue mineral. As you discovered in ITQ 16b, the sulphide mineral assemblages are characterized by different Cu : Fe and metal : sulphur ratios. For example, the chalcopyrite ($CuFeS_2$) – bornite ($Cu_5FeS_4$) zone has a high Cu : Fe ratio and a high metal : sulphur ratio, whereas the pyrite ($FeS_2$) zone has a Cu : Fe ratio of almost zero and a low metal : sulphur ratio. This reflects in part the different copper, iron and sulphide ion concentrations* in the fluids from which the minerals precipitated and it raises an important question. Did these zones result from a single surge of fluid that changed in chemical character as it migrated through the rocks?

*Strictly, these are the effective concentrations of the ions in solution, or *activities*.

Or did the zones result from discrete surges, each involving a fluid of different chemistry? Gustafson and Hunt were able to examine these two alternatives by studying field relationships and mineral textures.

Which possibility do *you* think is more likely?

It turned out that both processes were involved. Gustafson and Hunt found that two stages of mineralization took place corresponding to the early and late vein mineralization episodes. The early stage produced the central *chalcopyrite–bornite* zone and the surrounding *chalcopyrite–pyrite* zone. Superimposed on these zones, and produced by later mineralizing fluids, were the peripheral *pyrite* zone and the *pyrite–bornite* zone.

## 5.2   Secondary enrichment

In Section 2.1, we emphasized the importance of secondary enrichment in the evolution of porphyry copper deposits. As you know, El Salvador would not be an orebody were it not for this process. We shall now examine the secondary enrichment at El Salvador in more detail and start by looking at the ore minerals in polished section.

ITQ 18   Slides 4–6 are photomicrographs of the enriched ore viewed under reflected light. Slide 4 shows a field of view with the analyser out, and Slide 5, the same field of view with the analyser in. Slide 6 is viewed with the analyser out. Using Table 1 of the *Handbook*, Section XI, identify the minerals present in each slide. Which are primary minerals and which were formed by supergene processes?

ITQ 19   What textural evidence exists in these slides for the mode of formation of the supergene minerals?

The principal supergene sulphides at El Salvador are chalcocite, djurleite and digenite, three copper sulphides that have very similar chemical compositions and that are difficult to distinguish under the microscope (see Table 1 of the *Handbook*). Covellite, cuprite and even native copper are present in small amounts. As you may have discovered in Slides 4–5, these minerals form by replacement of hypogene minerals, mainly chalcopyrite and bornite; pyrite has not been replaced. The fluids involved in secondary enrichment also destroyed anhydrite in the upper levels of the deposit first converting it to gypsum ($CaSO_4.2H_2O$) and then removing it in solution. Gustafson and Hunt mapped the leached capping and the zone of secondary enrichment and plotted these on to the isometric section (Fig. 27).

There are several factors which can assist the secondary enrichment process:

(a)   *The neutralizing power* of the wall rock. If the host rock is one which reacts readily to acid attack, such as limestone, then supergene solutions will be rapidly neutralized. Little or no leaching and enrichment will then take place. On the other hand, relatively inert host rocks such as quartzite will do nothing to hinder leaching and enrichment.   **neutralizing power**

(b)   The *climate*. A warm climate and a deep, fluctuating water table are both important; hot, arid regions form the most favourable conditions.

(c)   The *amount of pyrite*. You already know from S26– that it is oxidation of sulphides to sulphate which makes the circulating groundwater acid and therefore able to leach copper from the rocks. A high proportion of pyrite in the leached rocks will therefore favour secondary enrichment.

Which of these factors helped the secondary enrichment process at El Salvador?

All these conditions were favourable at El Salvador: a relatively inert host rock of altered granodiorite; an arid climate; and a high concentration of pyrite in the upper levels (see Fig. 26). They combined to produce a large enrichment zone and an important orebody.

## 5.3   Wall-rock alteration

Wall-rock alteration is a metasomatic process (S23–, Block 4, Section 3.3.2)— a metamorphic process that involves a substantial change in the chemical   **wall-rock alteration**

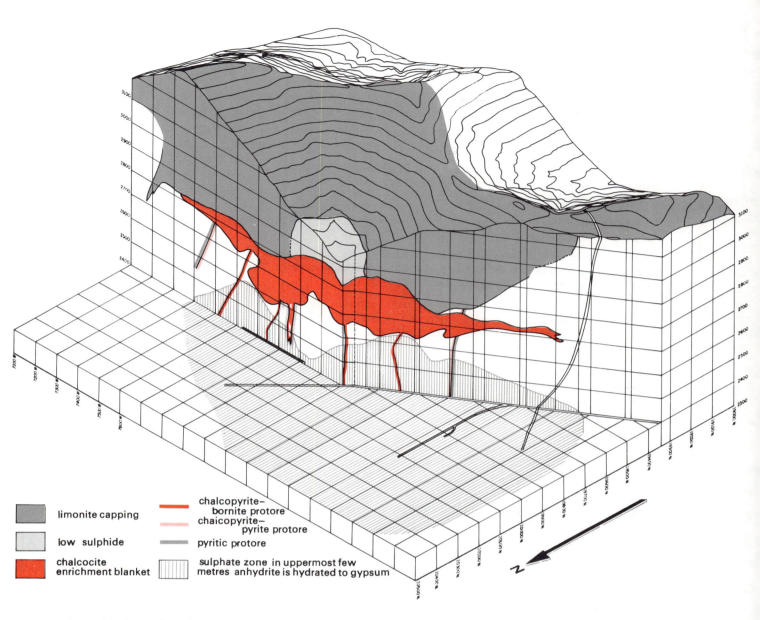

limonite capping

low sulphide

chalcocite
enrichment blanket

chalcopyrite–
bornite protore

chalcopyrite–
pyrite protore

pyritic protore

sulphate zone in uppermost few
metres anhydrite is hydrated to gypsum

*Figure 27* Isometric section through the El Salvador orebody showing the pattern of secondary enrichment (from Gustafson and Hunt, 1975).

composition of the rocks involved. It is produced within and around orebodies by the hydrothermal fluids that transport the ore constituents. During the migration of these fluids, the minerals in the surrounding rocks become unstable when they are suddenly subjected to a change in temperature and come into contact with a medium of different chemical composition. As a result, the chemistry, mineralogy and texture of the wall rocks change until they become stable under the new conditions. You have already encountered wall-rock alteration in this *Case Study*: the area of visible alteration showed up clearly on the aerial photograph of El Salvador (Fig. 16), alteration was recorded in the borehole log (Fig. 8), and in Section 3 we explained how alteration affects the behaviour of the rock during mining. Wall-rock alteration is, therefore, important in ore search and mining as well as ore genesis. Here, we shall concentrate on the pervasive alteration related to the disseminated mineralization, rather than on the less well-developed alteration halos around some of the veins. We shall begin by describing the different types of alteration and then attempt to explain some of the chemical and mineralogical changes which have taken place.

You will not be familiar with several of the minerals produced at El Salvador by wall-rock alteration. These are, however, minerals well worth knowing because they occur around many types of ore deposit, and are common constituents of regionally metamorphosed rocks. The minerals you will encounter in this Section are given in Table 11. You will need this Table *for reference only*.

### 5.3.1 Mineral assemblages

There are five distinct types of alteration described in the literature on ore deposits, each type characterized by a particular mineral assemblage. We shall explain each type in turn.

1 *K-silicate alteration* is characterized by the presence of new potash feldspar and new biotite. The potash feldspar is typically a perthite (S23–, *Reference Handbook for the Study of Minerals and Rocks*), and is formed by replacement of plagioclase feldspar, whereas biotite is formed by replacement of hornblende.

K-silicate alteration

2 *Propylitic alteration* is characterized by the breakdown of plagioclase and mafic minerals to chlorite and epidote, which gives the altered rock a greenish appearance.

propylitic alteration

3 *Sericitic alteration*, in which feldspar and mafic minerals are altered to a fine-grained mesh of the colourless-to-pale-green micaceous mineral, sericite.

sericitic alteration

4 *Argillic alteration* is distinguished by the presence of clay minerals, such as kaolinite, which have formed by alteration of feldspars and mafic minerals. The rock takes on a white, powdery appearance.

argillic alteration

5 *Advanced argillic alteration*, in which the main minerals are a complex series of aluminosilicates such as pyrophyllite ($Al_4Si_8O_{20}(OH)_4$), and even corundum ($Al_2O_3$) can be present. We shall not deal with this type of alteration, but you should be aware that it does exist.

advanced argillic alteration

All five alteration types are found at El Salvador. The recognition of alteration type and interpretation of the reactions that have taken place is best made by microscopic examination of thin sections. In Slides 7–12 we have provided photomicrographs of thin sections of four rocks from El Salvador for you to study.

| | | | |
|---|---|---|---|
| Rock A | Slide 7 (analyser out) | Rock C | Slide 10 (analyser in) |
| | Slide 8 (analyser in) | Rock D | Slide 11 (analyser out) |
| Rock B | Slide 9 (analyser in) | | Slide 12 (analyser in) |

TABLE 11   Composition, appearance and microscopic properties of some typical minerals produced by hydrothermal alteration.

| Mineral* | Composition | Usual occurrence** | Appearance in rock | Microscopic properties | |
|---|---|---|---|---|---|
| | | | | Polarized light | Crossed polars |
| Anhydrite | $CaSO_4$ | evaporites | pinkish to colourless crystals; sometimes fibrous | colourless, low relief | bright interference colours |
| Chlorite | $(Mg, Al, Fe)_6(Si, Al)_8O_{20}(OH)_4$ | clays, low-grade metamorphic rocks | green micaceous flakes | pleochroic (deep green–pale green) | anomalous (often blue or brown) interference colours |
| Epidote | $Ca_3Al_2Si_3O_{12}(OH)$ | low to medium grade metamorphic rocks | yellow-green crystals | yellow-green | bright interference colours |
| Kaolinite | $Al_4Si_4O_{10}(OH)_8$ | clays, weathered rocks | very fine grained, white clay mineral | pale brown mesh of crystals | almost isotropic |
| Perthite | $(K, Na)AlSi_3O_8$ | high grade metamorphic rocks, granites | colourless to pink-orange crystals | colourless, low relief | shows a complex exsolution intergrowth of sodic and potassic feldspar |
| Sericite | $KAl_2AlSi_3O_{10}(OH)_2$ | metamorphic rocks | very fine grained; micaceous; white to pale green | colourless, low relief | bright interference colours |

* You will also encounter in the text the minerals *andalusite* ($Al_2SiO_5$) and *pyrophyllite* ($Al_4Si_8O_{20}(OH)_4$), as well as *biotite*, *plagioclase feldspar*, and *quartz* with which you will already be familiar.

** Of course, these minerals also occur around ore deposits!

**ITQ 20** Use Slides 7–12 and Table 11 to answer the following questions.

(a) *For rock A* The phenocryst has been completely replaced during the alteration process. What is the phenocryst and what has replaced it? What type of alteration is this?

(b) *For Rock C* What is the phenocryst and what mineral is starting to replace it? What type of alteration is this?

(c) *For rock D* What are the green minerals in this altered andesite? What type of alteration is this?

On the basis of microscopic examination of large numbers of samples, Gustafson and Hunt were able to plot the type of wall rock on mine sections, and finally on the isometric section (Fig. 28). They observed a zoning of alteration type that could be correlated with the zoning of mineralization.

zoning of wall-rock alteration

**ITQ 21** Compare Figure 28 with the mineralization zoning patterns in Figures 26 and 27:

(a) Which alteration zones are spatially related to the early mineralization zones of chalcopyrite–pyrite and chalcopyrite–bornite?

(b) Which alteration zone is related to the pyrite zone of the later mineralization episode?

(c) Which alteration zone is related to the zone of secondary enrichment?

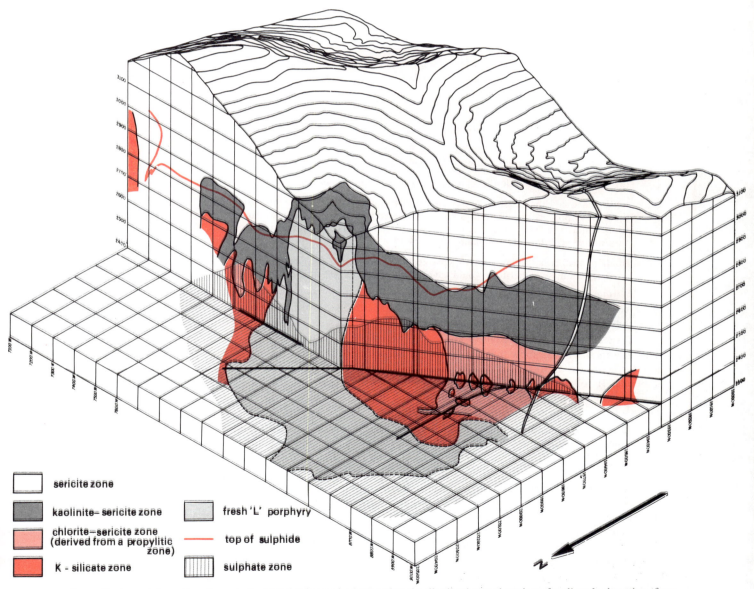

sericite zone

kaolinite–sericite zone

chlorite–sericite zone (derived from a propylitic zone)

K - silicate zone

fresh 'L' porphyry

top of sulphide

sulphate zone

*Figure 28* Isometric section through the El Salvador orebody showing the distribution and zoning of wall-rock alteration (from Gustafson and Hunt, 1975).

This correlation between mineralization and alteration zoning should not surprise you. It was, after all, the same fluids that produced the mineralization which also caused the wall-rock alteration. To summarize then: the early mineralization episode produced a central K-silicate alteration zone with a peripheral propylitic zone; the later mineralization episode produced sericitic alteration; whereas the secondary enrichment episode produced a zone of argillic alteration.

### 5.3.2  The alteration process

We can illustrate the principles involved in wall-rock alteration by considering the alteration of porphyritic granodiorite at El Salvador. We cannot investigate propylitic alteration at this point because it occurs in altered andesite, not granodiorite, but we can examine the other alteration types. During the early stage of mineralization, K-silicate alteration took place, giving a mineral assemblage dominated by K-feldspar, with biotite and quartz. During the late stage of mineralization this rock underwent sericitic alteration producing an assemblage dominated by sericite, with chlorite and quartz Finally, during secondary enrichment, the argillic alteration assemblage of kaolinite accompanied by sericite, chlorite and quartz was produced. We want to know under what conditions K-feldspar, sericite and kaolinite become the dominant mineral phase. Two chemical reactions are important:

chemical changes during wall-rock alteration

$$3KAlSi_3O_8 \ + \ 2H^+ \rightleftharpoons KAl_2AlSi_3O_{10}(OH)_2 \ + \ 6SiO_2 \ + \ 2K^+ \qquad (1)$$

potash feldspar | supplied by the fluid | sericite | quartz | removed in the fluid

$$4KAl_2AlSi_3O_{10}(OH)_2 \ + \ 6H_2O \ + \ 4H^+ \rightleftharpoons 3Al_4Si_4O_{10}(OH)_8 \ + \ 4K^+ \qquad (2)$$

sericite | supplied by the fluid | kaolinite | removed in the fluid

**ITQ 22**  Now examine these two chemical reactions.

(a) What element is being introduced into the system and what is being removed, assuming the reactions are going to the right?

(b) Which type of alteration is represented by each of these equations 1 and 2?

(c) We have said that these reactions are reversible, but in reality they move to the right to produce the alterations you have named in (b). In order for this to happen, must the fluids have a high or low $K^+/H^+$ ratio, and does this mean that they will be acid (low pH) or alkaline (high pH)?

(d) If hydrothermal fluids were to leave feldspar unaffected, can you deduce from equation 1 whether they would have a high or low $K^+/H^+$ ratio?

(e) Fluids do not simply carry around positive ions in isolation; there must be a corresponding number of negative ions to balance the charges. Which negative ion is most likely to be present in the hydrothermal fluids?

The answers to ITQ 22 show that acid hydrothermal solutions with low pH and low $K^+/H^+$ ratios are necessary for the breakdown of feldspar to sericite or to kaolinite. It follows that the stability of these minerals in the hydrothermal environment will depend on the $K^+/H^+$ ratio in the fluids.

Is that the only factor involved, or will the stability of these minerals be dependent upon other things too? If so, what additional variable do you think is most important?

The very name *hydrothermal* fluid suggests the answer. The other important variable must be temperature. It follows then that if we were to plot a graph of $K^+/H^+$ against temperature for hydrothermal fluids, we could delineate fields in which the minerals K-feldspar, kaolinite and sericite are stable. Figure 29 is just such a graph; it is simply a phase diagram of the kind you have studied in other Courses, but with slightly different parameters involved.

temperature–composition phase diagram

How do you think that diagrams such as Figure 29 are compiled—can they be compiled from measurements taken directly in the field?

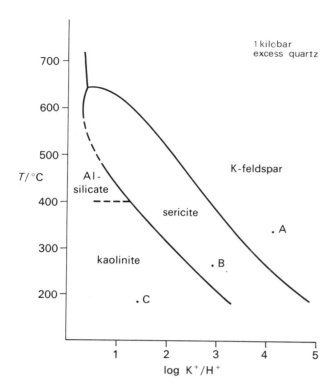

Figure 29 Stability relations of kaolinite, sericite and K-feldspar in chloride solutions with respect to temperature and KCl/HCl ratio (from Rose, 1971).

Well, no, in general they cannot. Although it is true that there are areas where hydrothermal fluids can be sampled (an obvious example is the Salton Sea; see S26–, Block 3, Section 3.5.2), graphs of this type are compiled from laboratory measurements on mineral stabilities. They can be *tested* by field observations, but not actually compiled in this way.

ITQ 23 You will recall from ITQ 21 that the K-feldspar, sericitic and argillic alterations took place in three stages at El Salvador: an *early* mineralization stage, a *late* mineralization stage, and a *secondary enrichment* stage.

(a) What does Figure 29 tell you about the temperature and $K^+/H^+$ ratios characterizing the mineralizing fluids?

(b) What ore minerals were probably deposited from fluids approximating to points A, B and C on Figure 29?

So far, then, we have explained how the sericitic, the argillic and the K-silicate alterations at El Salvador involved fluids of different temperature and/or composition. But there is also another factor that must be taken into consideration. As a fluid migrates through the rock, effecting wall-rock alteration, *its chemistry, pressure and temperature will be continually changing*. With this in mind, let us examine in more detail the early alteration at El Salvador and discover why the type of alteration changed from K-silicate at the centre of the orebody to propylitic at the periphery. The fluids were initially producing K-silicate alteration by taking part in the following reactions:

$$\text{plagioclase feldspar} \;+\; \underset{\substack{\text{supplied}\\\text{by the fluid}}}{K^+} \;\rightleftharpoons\; \text{potash feldspar} \;+\; \underset{\substack{\text{removed}\\\text{in the fluid}}}{(Na^+, Ca^{2+})} \qquad (3)*$$

$$\text{hornblende} \;+\; \underset{\text{supplied by the fluid}}{(H^+, Mg^{2+}, K^+)} \rightleftharpoons \text{biotite} \;+\; \underset{\substack{\text{removed}\\\text{in the fluid}}}{(Na^+, Ca^{2+})} \qquad (4)*$$

ITQ 24 As the fluids migrate through the rock, do you think that the following parameters might increase, decrease or remain unchanged?

1 pressure      3 the $K^+/H^+$ ratio      5 the $K^+/Na^+$ ratio

2 temperature      4 the $K^+/Ca^{2+}$ ratio

*See footnote on p. 48

47

The implication of your answers to ITQ 24 is that the fluids may not remain in the K-feldspar stability field. Firstly, if temperature and $K^+/H^+$ ratio both decrease, the fluid will move towards the sericite phase field in Figure 29, and in some cases sericitic alteration may eventually take place. Secondly, and this is observed at El Salvador, the decrease in $K^+/H^+$ and $H^+/Na^+$ ratios enables reactions such as the following to take place:

$$\text{biotite} + H^+ \rightleftharpoons \text{chlorite} + \text{quartz} + K^+ \qquad (5)^*$$
$$\underset{\substack{\text{supplied} \\ \text{by the fluid}}}{} \qquad \underset{\substack{\text{removed} \\ \text{in the fluid}}}{}$$

$$\text{plagioclase feldspar} + \text{quartz} + H_2O + Na^+ \rightleftharpoons \text{epidote} + \text{albite} + H^+ \qquad (6)^*$$
$$\underset{\substack{\text{supplied} \\ \text{by the fluid}}}{} \qquad \underset{\substack{\text{removed} \\ \text{in the fluid}}}{}$$

So what type of alteration is that?

The epidote and chlorite are characteristic minerals of *propylitic* alteration, which occurs outside the K-silicate alteration zone at El Salvador.


## 6.0 El Salvador: fluid inclusion and isotope studies

**Study comment** You probably did not realize until now just how complicated the history of an ore deposit can be! In the last Section you read about early, transitional and late inputs of mineralizing fluids, each of which caused its own pattern of mineralization and wall-rock alteration; and superimposed on this were the effects of leaching and enrichment by supergene fluids. But to the geologist studying the origin of ore deposits this information is not enough. He wants to know where these fluids have come from and what their temperatures and salinities were. For this he must take samples back to the laboratory to carry out fluid inclusion and stable isotope studies. When you have read this Section, and *Handbook* Section VI, *Stable isotopes*, you should be aware of how such studies are carried out, how the data are interpreted, and the type of information that the geologist expects to obtain.


### 6.1 Fluid inclusions

Fluid inclusions are the best samples available of the fluids that caused mineralization and alteration, and their study is an integral part of an investigation into the origin of an ore deposit. The inclusions provide information that can be used to interpret the temperature and the composition of the mineralizing fluids. Fluid inclusions are small quantities of fluids that have been trapped in the ore and gangue crystals (S26–, Block 3, Section 3.5.2). Most fluid inclusions are less than 0.1 mm across and can only be seen under the microscope. Inclusions larger than 0.1 mm are rare, although some museums do have examples several centimetres across in their collections. The inclusions may be *primary* (formed at the time of growth of the immediately surrounding parts of the crystal) or *secondary* (introduced after the crystal has grown, perhaps along a fracture). If the fluid is introduced into the centre of the crystal while the outside of the crystal is still growing, the inclusion is termed *pseudosecondary*. Geologists are most interested in primary inclusions, since these are very likely to represent samples of the fluid actually involved in the mineralization. However, in some rocks it can be very difficult to identify which inclusions are primary, and in others no primary inclusions may exist.

primary, secondary and pseudo-secondary fluid inclusions

Fluid inclusions can contain several phases in various proportions. They can contain a second immiscible liquid such as oil, a bubble of gas or vapour, or one or more precipitated solids such as halite (NaCl). Most vapour and solid phases have separated out from a homogeneous liquid as it cooled. Vapour bubbles result when the liquid, which filled the inclusion at high temperatures, shrinks in volume on cooling. Solid phases form when the liquid becomes saturated in these phases as it cools, and crystals, known as *daughter minerals*, nucleate within the inclusions.

daughter minerals

*These partial reactions, and those in equations 3 and 4, have not been balanced because they involve minerals of variable composition. The general compositions of these minerals are given in Table 11 and in S23–(*Reference Handbook for the Study of Rocks and Minerals*).

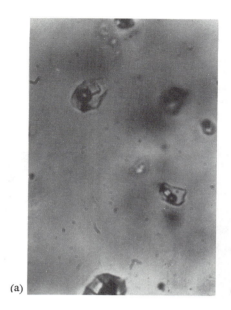
(a)

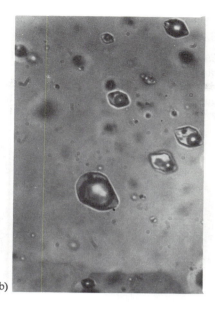
(b)

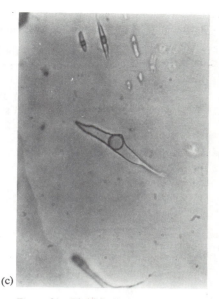
(c)

*Figure 30* Fluid inclusions at El Salvador. Figure 30a shows typical type I inclusions, Figure 30b type II inclusions and Figure 30c type III inclusions.

microscope heating stage

**ITQ 25**  At El Salvador, Gustafson and Hunt recognized three distinct types of fluid inclusion, which they termed types I, II and III. These are illustrated in Figure 30. How many phases are present in each type of inclusion?

Gustafson and Hunt carried out three experiments on these inclusions: heating, freezing and crushing. Heating experiments are carried out on a *heating stage*, a piece of apparatus which fits on to the microscope stage and which simultaneously heats the sample, takes its temperature and enables the sample to be viewed under the microscope (Fig. 31). The sample itself is prepared as a thin polished slice which will heat up rapidly and be easy to view. When a fluid

*Figure 31* A picture of the heating stage used to study fluid inclusions at the Royal School of Mines in London. The leads to the heating stage are used for the heating element and a thermocouple. The output from the thermocouple is printed out on a chart recorder.

inclusion containing a vapour bubble is heated on this apparatus, the bubble becomes smaller and smaller until, at some temperature, it disappears (Fig. 32).

This temperature, known as the *homogenization temperature*, represents the true temperature of trapping of the liquid, provided that the liquid was at its boiling point when it was trapped. Under these circumstances, the slightest shrinkage would have produced some vapour. If, however, when the liquid was trapped it was under sufficiently high pressure to prevent its boiling, it would have to

homogenization temperature

*Figure 32* The effect of heating on a fluid inclusion. The gas bubble disappears at 210 °C, the homogenization temperature (from Roedder, 1962).

*Figure 33* The effect of freezing on a fluid inclusion. The inclusion is frozen solid at −29.5 °C. It starts to melt at −29 °C (its 'first melting temperature') and it is completely molten at −3.15 °C (its freezing temperature) (from Roedder, 1962).

cool to a lower temperature before the vapour began to form. So, under these circumstances, the heating experiment will register only this lower temperature. This can be corrected to the true temperature of formation, provided that we know, from geological reasoning, the total pressure existing at the time of trapping. Figure 34 shows the sort of calibration curve that can be used to make this correction.

pressure correction

> **ITQ 26** Type III fluid inclusions in one sample of quartz from El Salvador homogenized at 200 °C. Freezing studies suggested that they had a salinity of 10 per cent equivalent NaCl.* Assuming an overburden pressure of 1 kb, what was the true temperature of trapping of the fluid? (1 kb = 1 000 atm = $10^5$ Pa.)

Only type III inclusions at El Salvador required this pressure correction. Some of the type I and II inclusions contained more vapour than liquid, evidence that the liquid was boiling. If you look carefully at Figure 30b you should notice one such inclusion—it is in the centre of the photograph.

Heating experiments also serve to identify solid phases. For example, a crystal of sylvite (KCl) will dissolve eight times faster than a crystal of halite on heating.

*Salinities are usually expressed as percentages (by weight) of an equivalent NaCl content (i.e. the total salinity, if all the salts present are assumed to be NaCl).

Freezing experiments are carried out on a freezing stage, which is similar to a heating stage but with a mechanism involving liquid nitrogen for *cooling* the sample. Since salts depress the freezing point of water, the depression of the freezing point of the liquid below 0 °C is used to measure its salinity (Fig. 33).

**microscope freezing stage**

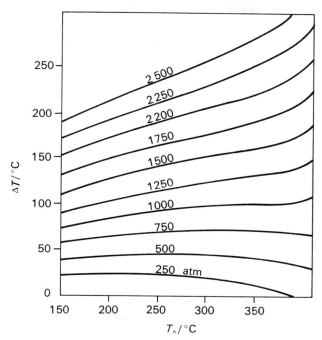

*Figure 34* Temperature correction ($\Delta T$) to be applied to the homogenization temperature ($T_h$) for different pressures of trapping of a 10 per cent NaCl solution (from Klevtsov and Lemmlein, 1959).

The crushing experiment involves crushing small mineral grains individually under oil. This is carried out under the microscope using special apparatus—a microscope crushing stage. This causes the compressed gas (usually $CO_2$) to be released as a bubble. The size of this bubble when it is released, and its subsequent expansion or contraction in the oil, give a rough indication of the proportion of $CO_2$ in the inclusion and its vapour pressure.

**microscope crushing stage**

These experiments are only a fairly small proportion of the total number of experiments that can be carried out on fluid inclusions, but they are among the most important. Recently, for instance, methods have been devised for extracting the fluid and subjecting it to various analytical tests.

Gustafson and Hunt studied the fluid inclusions from quartz grains within the *A*-, *B*- and *D*-veins. *A*- and *B*-veins both contained type I, type II and type III inclusions (Fig. 30), but *D*-veins contained only type III inclusions. Using the techniques described above, they were able to identify the properties of each of these types of inclusion.

> *Type I inclusions* homogenized over the temperature range 360–600 °C. The heating behaviour of single inclusions containing halite and sylvite daughter minerals confirmed visual estimates that the fluids contained 35–40 per cent NaCl and 12 per cent KCl; freezing experiments were, however, unsuccessful. When type I inclusions were crushed in oil, the bubble released collapsed immediately suggesting that the vapour pressure was very low.
>
> *Type II inclusions* also homogenized over the temperature range 360–600 °C. Again, the freezing experiments were unsuccessful although the salinity was presumed to be low. The crushing experiment suggested a $CO_2$ content of about $\frac{1}{8}$ to 8 atmospheres at room temperature; using this figure, order of magnitude calculations put the fluid as containing 4 per cent $CO_2$ by weight, with a pH of 4.
>
> *Type III inclusions* homogenized between 175 and 310 °C. Their freezing behaviour indicated an equivalent NaCl content of 5 to 20 per cent. The crushing experiment was not successful.

**ITQ 27** As Gustafson and Hunt pointed out, these inclusions must have been trapped over a range of temperatures and at different times. It is not certain which, if any, inclusions are actually primary. However, there does seem to have been a change in the character of fluids trapped during the evolution of the vein system. So what can you say about the temperature and salinity of:

(a) Early and transitional mineralizing fluids?

(b) Late mineralizing fluids?

Finally, it would be inappropriate to teach anything about fluid inclusions without stating some of the many problems involved in the subject. The exact nature of the fluid is perhaps the major uncertainty. It is often difficult to recognize primary inclusions; even when we can, we may still not know whether the trapped fluid is actually representative of a fluid present at the time of formation of the inclusion or whether there has been leakage in or out of the inclusion after trapping. In making temperature calculations, the pressure on trapping is often poorly known; the results have large errors because the samples are so small, and ranges rather than precise values are obtained. However, despite these problems, the studies on El Salvador gave useful results, as you will discover in the next Section.

## 6.2  Stable isotopes

You should by now know something about the principles and application of oxygen and hydrogen isotopes, from both the *Handbook*, Section VI, *Stable isotopes*, and the *Igneous Case Study*.

*If you are still not clear about the principles and applications of stable-isotope geochemistry, reread the Handbook, Section VI, now, paying particular attention to Section 2.2.*

Imagine, then, that you intend to use oxygen and hydrogen isotopes to investigate the alteration and mineralization at El Salvador.

  1  What rock types would you want to study most?

  2  Where would you get the *water* for isotopic analysis?

  3  What information would the results supply you with?

The actual study was carried out by Sheppard and Gustafson. They collected a representative suite of rocks from the different zones of mineralization—from the various porphyritic intrusions, from the *A*-, *B*- and *D*-veins and from the zone of secondary enrichment. The water came from the hydrous minerals in

TABLE 12  Hydrogen and oxygen isotope analyses of minerals from El Salvador (from Sheppard and Gustafson, in press).

| Sample No. | Sample | Mineral | $\delta^{18}O\%_0$min | $\delta D\%_0$min |
|---|---|---|---|---|
| ES2689 | *L*-porphyry | Plagioclase | 7.1 | −85 |
| | | Biotite* | 4.5 | |
| | | Hornblende* | 5.9 | −73 |
| 2691 | *L*-porphyry | Plagioclase | 7.6 | |
| | | Biotite* | 5.0 | −78 |
| 2699R | *X*-porphyry | Quartz | 9.2 | |
| | | K-feldspar | 8.3 | |
| | | Plagioclase | 8.2 | |
| | | Biotite* | 4.3 | −73±1(2) |
| 1910R | Andesite | Biotite | 4.0 | −75 |
| | *A*-vein | Quartz | 9.3±0.1(2) | |
| | | K-feldspar | 8.5 | |
| 4304 | *A*-vein | Quartz | 9.4 | |
| | | K-feldspar | 9.9 | |
| 7536 | *B*-vein | Quartz | 10.3 | |
| 2699V | *B*-vein | Quartz | 9.3±0.2(2) | |
| 1116 | *X*-porphyry | Quartz | 9.7 | |
| | | Sericite* | 6.9±0.1(2) | −59 |
| | | Chlorite* | 4.5 | −65 |
| 7576V | *D*-vein | Quartz | 12.0±0.1(2) | |
| 7576H | *D*-vein halo | Quartz | 10.7±0.1(2) | |
| | | Sericite* | 9.3±0.0(2) | −68 |
| 1417 | *D*-vein halo | Quartz | 9.2 | |
| | | Sericite | 8.2 | −78±1(2) |
| | | Kaolinite* | 9.4 | −79 |
| 7486 | *D*-vein | Quartz | 10.9 | |

*$H_2O$ concentrations in hydrous minerals were also determined by Sheppard and Gustafson since this affects isotope fractionation.

the rocks (hornblende, biotite, sericite, chlorite, kaolinite and pyrophyllite) which were separated and analysed for D/H and $^{18}O/^{16}O$ ratios; non-hydrous minerals such as quartz and feldspar were analysed for $^{18}O/^{16}O$ only. The actual data, taken from Sheppard and Gustafson (1960) are reproduced in Table 12. They used these data to investigate the source of the fluids involved in the mineralization and alteration, and to estimate the temperature of these fluids. Here, we present their interpretations and follow through the calculations and reasoning behind them.

### 6.2.1  A stable isotope geothermometer

Look at Table 12. You will notice that several minerals have been analysed from each rock. Furthermore, the various minerals within any one rock have different $\delta^{18}O$ (and $\delta D$) values. In *Handbook* VI, Section 2.2, we explained how the observed difference in $\delta^{18}O$ values between any pair of coexisting minerals is related to the temperature at which isotopic exchange took place between these minerals and the mineralizing fluids. Figure VI–2 in the *Handbook* shows this relationship for the various mineral pairs analysed by Sheppard and Gustafson. You can use this diagram to work out the temperatures of the fluids associated with some of the intrusive rocks from El Salvador.

stable isotope geothermometry

> **ITQ 28**  Isotopic data for several minerals is available for rocks 2689, 2691 and 2699. Complete Table 13 using $\delta^{18}O$ values from Table 12 and the calibration curves from Figure VI–2 in the *Handbook*. What assumptions have you made?

TABLE 13   Fluid temperature calculated from oxygen isotope compositions of coexisting mineral pairs (to be completed as ITQ 28).

|  | Sample No. 2689 | 2691 | 2699 |  | Sample No. 2689 | 2691 | 2699 |
|---|---|---|---|---|---|---|---|
| $\Delta$(qtz—bi) |  | — | — | $T=$ |  | — | — |
| $\Delta$(pl—bi) |  |  |  | $T=$ |  |  |  |
| $\Delta$(ksp—bi) |  | — | — | $T=$ |  | — | — |

Key: qtz=quartz; bi=biotite; ksp=K-feldspar; $\Delta$(qtz—bi)=$\delta^{18}O_{qtz}-\delta^{18}O_{bi}$ etc.

The temperatures you obtained in ITQ 28 are of the right order of magnitude for fluids in contact with a molten porphyritic granodiorite magma. The similarity between the temperatures determined from different mineral pairs suggests that equilibrium may have been reached. However, the mineral pairs analysed in the other rocks do not give reasonable temperatures. This is probably because equilibrium conditions were not strictly met. Small departures from equilibrium, which do not significantly affect the studies in Section 6.2.2, can invalidate the use of stable isotopes as a geothermometer. There are also other problems in dealing with these altered rocks. For example, several generations of quartz may exist within one rock.

### 6.2.2  Origin of the fluids

You should remember from the *Handbook*, Section VI, that the $\delta D$ versus $\delta^{18}O$ diagram provides a method of studying the origin of mineralizing fluids. By plotting the isotopic composition of the fluids associated with the various rock types, it should be possible to work out their source.

> What types of water do you think are most likely to be present?

The choice is most probably between *magmatic water* derived from the igneous intrusions, and *meteoric water* that has penetrated into the zone of mineralization along faults and fractures. Table 12 is no use to us as it is, because the isotopic data are given for minerals in equilibrium with the fluids and not for the fluids themselves.

> Can you think of any way to convert isotopic data on the minerals to isotopic data on the associated fluids?

You can use Figure VI–1 in the *Handbook*, which shows that the difference in $\delta^{18}O$ between any mineral and the associated fluid is a function of temperature. Knowing $\delta^{18}O$ for the mineral, and knowing the temperature at which exchange took place (from fluid inclusions, isotopes or other geological reasoning), $\delta^{18}O_{H_2O}$ can be worked out using this diagram. The recalculated data are given in Table 14.

TABLE 14  Calculated $\delta^{18}O$ and $\delta D$ values of the hydrothermal fluids involved in forming the El Salvador deposit (from Sheppard and Gustafson, in press).

|  | Sample | Temperature* °C | Mineral | $\delta^{18}O_{H_2O}$** | $\delta D_{H_2O}$** |
|---|---|---|---|---|---|
| 2689 | *L*-porphyry | 750–625 | Plagioclase | 7.7±0.4 | |
|  |  |  | Biotite | 7.3±0.1 | −62±4 |
|  |  |  | Hornblende | 7.5±0.4 | −55±4 |
| 2691 | *L*-porphyry | 750–625 | Plagioclase | 8.2±0.4 | |
|  |  |  | Biotite | 7.8±0.1 | −57±4 |
| 2699 | *X*-porphyry | 650–520 | Quartz | (a) | |
|  |  |  | Biotite | 6.9±0.1 | −45±5 |
| 1910 | Andesite | 650–520 | Biotite | 6.6±0.1 | −47±5 |
| 1910 | *A*-vein | 650–520 | Quartz | 7.3±0.8 | |
|  |  |  | K-feldspar | 7.9±0.7 | |
| 4804 | *A*-vein | 650–520 | Quartz | 7.4±0.8 | |
|  |  |  | K-feldspar | 9.2±0.7 | |
| 7536 | *B*-vein | 450–350 | Quartz | (b) | |
| 2699 | *B*-vein | 450–350 | Quartz | 3.8±1.4 | |
| 1116 | *X*-porphyry | 400–300 | Quartz | 4.2±1.4 | |
|  |  |  | Sericite | 5.0±0.8 | −32±5 |
|  |  |  | Chlorite | 5.2±0.8 | −25±5 |
| 7576 | *D*-vein | 350–200 | Quartz | (c) | |
| 7576 | *D*-vein Halo | 350–200 | Quartz | −0.1±3.8 | |
|  |  |  | Sericite | 5.6±1.9 | −33±5 |
| 1417 | *D*-vein Halo | 350–200 | Sericite | 4.5±1.9 | −43±5 |
|  |  |  | Kaolinite | 6.1±1.5 | −44±5 |
| 7486 | *L*-vein | 350–200 | Quartz | 0.1±3.8 | |

*From fluid inclusion and isotopic geothermometry.

**The mean $\delta$-value is given; the range is derived from the temperature limits

> ITQ 29  To check that you understand how to convert $\delta^{18}O$ for a mineral to $\delta^{18}O$ for the mineralizing fluid in contact with it, fill in the three spaces (a), (b) and (c) in Table 14.

Now that the data are at last in the right form, you should be able to use them to work out the origin of the fluids involved in the primary mineralization. Figure 35 gives you the base for plotting the $\delta D_{H_2O}$ and $\delta^{18}O_{H_2O}$ data, and shows the meteoric water line, SMOW, and the field of primary magmatic water. The one further piece of information you need, the approximate position of the El Salvador meteoric water on the meteoric water line, has been estimated by Sheppard and Gustafson and is also plotted on this diagram.

> ITQ 30  Plot the data from Table 14 on to Figure 35.
>
> (a) Locate the analyses most likely to have had a magmatic origin. Where do these plot?
>
> Did the fluids associated with the *D*-veins have a similar origin?

While you were doing ITQ 29, you may have realized that the meteoric water that percolates to deep levels does not stay on the meteoric water line. The isotopic composition of the water gets modified within the Earth in the same way as geothermal waters and formation waters (see *Handbook*). Isotopic exchange takes place between the meteoric water and the rocks it passes through and, as a result, the $\delta^{18}O$ (and to a much lesser extent $\delta D$) in the water moves to higher values. So we cannot tell from Figure 35 the relative amounts of magmatic and meteoric water associated with the *D*-veins, although we can say that some meteoric water is involved.

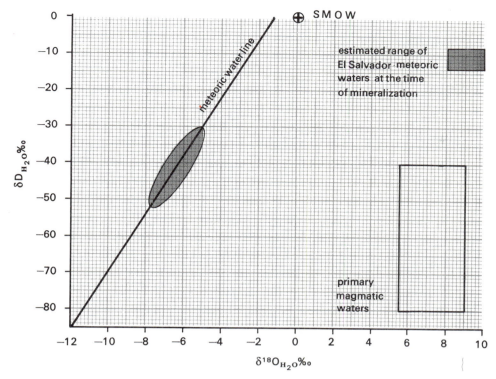

Figure 35  δD – δ¹⁸O graph showing SMOW, the meteoric water line and the field of primary magmatic water (for use with ITQ 30).

Another way of approaching the problem is to plot the isotopic data (e.g. $\delta^{18}O$) against temperature. In this case you will expect magmatic waters to have higher $\delta^{18}O$ values and higher temperatures than meteoric waters. During mixing, both $\delta^{18}O$ and temperatures should fall. Sheppard and Gustafson's plot of $\delta^{18}O_{H_2O}$ against mean fluid temperature is reproduced in Figure 36, using the data given in Table 14.

Does this diagram support the evidence from ITQ 30, that meteoric water was involved in the formation of the D-veins?

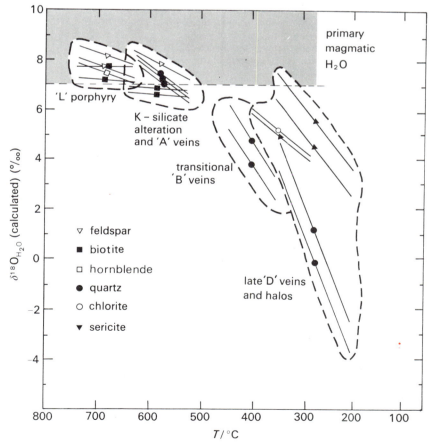

**δ¹⁸O-temperature diagram**

Figure 36  δ¹⁸O – temperature graph for El Salvador samples (from Sheppard and Gustafson, in press).

The decrease in $\delta^{18}O$ values and fluid temperatures from $A$-veins to $D$-veins suggests that the meteoric water component increases from $A$-veins (in which it is probably very small) to $D$-veins (in which it is probably the major component). The low salinity inclusions in $D$-veins could be explained in terms of dilution of a high-salinity magmatic water component by a low-salinity meteoric water component.

The conclusions we can make, therefore, are that the fluids associated with Early mineralization and alteration ($A$-veins) are primarily magmatic in origin and are derived from the porphyritic intrusive rocks; the fluids associated with Transitional mineralization ($B$-veins) contain some meteoric water; and the fluids associated with Late mineralization and alteration ($D$-veins) contain a large proportion of meteoric water that has penetrated deeply and mixed with magmatic water.

### 6.2.3 Sulphur isotopes

Sulphur has stable isotopes, $^{32}S$ and $^{34}S$, and these are used in economic geology to study the origin of sulphur in the mineralizing fluids. We have deliberately avoided this subject so far, because sulphur isotope data are much more difficult to understand and interpret than oxygen and hydrogen isotope data. The main reason is that there are many more variables—oxygen and sulphur partial pressures in the fluid, the pH and temperature of the fluid, and whether the sulphur exists as $HS^-$, $S^{2-}$, or $SO_4^{2-}$, for instance. Gustafson and Field did analyse for sulphur isotopes in the sulphides and in anhydrite at El Salvador. Briefly, their tentative conclusions were that the sulphur associated with Early mineralization gave results consistent with a magmatic source, whereas the sulphur associated with Late mineralization had a different source. It could have been derived from sulphides and anhydrite that were produced during Early mineralization and that were dissolved at depth and redeposited at higher levels. Alternatively, it could have been leached from surrounding volcanic rocks by circulating meteoric waters and then redeposited as sulphides.

**sulphur isotope studies**

## 7.0 The genesis of the El Salvador deposit

**Study comment** The aim of this Section is 'synthesis'. We shall try to show you how to build up a model of the genesis of the ore deposit, drawing upon *all* the information you have accumulated from Sections 3–6. We hope that once you have completed this Section you will be able to synthesize data on *any* hydrothermal ore deposit to give a comprehensive and structured account of its genesis.

### 7.1 Ideas on ore genesis

Look at Figure 15 in Block 3 of S26–. It shows a vein rich in chalcopyrite. Ideas on the origin of such a vein have changed throughout history. The Greek and Roman philosophers would have explained such a vein as having been exhaled from an animated Earth, or that it was part of a subterranean plant or tree whose roots lay deep within the Earth. The alchemists of the sixteenth and seventeenth centuries would have believed that this vein was generated by the Sun or the planets. If you read Latin and had lived in the sixteenth century you might have come across the works of Georg Bauer, who was known by his Latin name of Georgius Agricola. His work *De Re Metallica*, published in 1556, is the foundation of modern economic geology. Agricola would have explained that the ore was deposited from circulating solutions in a channel, that was younger than the surrounding country rock. In the seventeenth century, Niels Stensen, usually known as Steno, would have explained that the ore formed from vapours ascending open fissures; whereas in the eighteenth century a popular theory advanced by Charpentier proposed that such ores were formed by the alteration of country rock.

In the late eighteenth and early nineteenth centuries, geology was dominated by the diametrically conflicting views of James Hutton (a Scot) and Abraham Gottlob Werner (a German). Hutton, the 'Plutonist', believed that igneous rocks and ore deposits (including our chalcopyrite vein) were derived from molten magmas at depth and injected into their present position. Werner, the 'Neptunist', believed that basalts, sedimentary rocks and ore deposits were laid down as sediments in a primaeval ocean. According to this viewpoint, the chalcopyrite vein would be made of sediment filling a crack in the floor of this ocean. In the early twentieth century the great economic geologist, Waldemar Lindgren, would have classified the vein as a hydrothermal deposit formed by precipitation from a hot aqueous solution. He would further have classified the deposit in terms of temperature and depth of formation as hypothermal (300–500 °C), mesothermal (150–300 °C) or epithermal (50–150 °C).

Lindgren's classification of hydrothermal deposits

You should already, from S26–, recognize the vein as a hydrothermal ore. After you have completed this *Case Study* you should also be able to use data from isotopic, fluid inclusion, geochemical, and geological studies to build up a complex model for the genesis of the vein. But do not necessarily imagine that you will arrive at the complete answer. The ancient Greeks thought they had the answer too! Bear in mind, therefore, this quotation from the mining geologist Ira Joralemon (1930):

> The geologist must not devote himself to any one theory or the facts that support it long enough to fall hopelessly in love with it. He must make each theory the object of a summer flirtation and not a wife—and he must be ready to throw each one over the moment a more attractive mental maiden comes along.

**ITQ 31** Classify the approach to ore genesis used by the ancient Greeks, the alchemists, Werner, Hutton, Lindgren and yourself in terms of (a) inductive, (b) deductive, (c) hypothetico-deductive reasoning.

## 7.2 Causes of ore deposition

Before we can begin to synthesize the El Salvador data, it is necessary to know something about the factors which can cause ore deposits to form. We can explain ore deposition in terms of structural controls and chemical controls. The structure and textures of rocks control the migration and concentration of the ore-bearing fluids, whereas chemical changes in the ore fluids cause deposition of ore minerals, whether by crystallization within cavities in the host rock or by replacement of the minerals within the rock. Let us consider the structural controls first. In many respects, ore fluids behave like oil or gas. They must migrate from their source to the place of ore deposition through passageways in the rock. If the ore fluids are collected in a 'trap' analogous to the oil trap this will improve the possibility of ore deposition.

structural controls on ore deposition

*Now read Sections 4.3.1 and 4.3.2 of S26–, Block 2.* This explains the importance of permeable rocks as a path for ore fluids.

**ITQ 32** Which of the following rock types are most likely to be permeable to ore fluids?

(a) shale  (b) conglomerate  (c) granite

(d) reef limestone  (e) quartzite  (f) beach sand

But the permeability is by no means the most important factor, particularly at great depth, where permeabilities are mostly very low.

> Can you suggest, from your knowledge of field relations, other structures which will act as channels for the fluids?

The most obvious are joints, folds and faults, although features such as bedding planes and unconformities are also important. Some fractures act as long continuous channelways and are a very efficient means of transporting the large volumes of fluid needed to produce an ore deposit. Many ore deposits are spatially related to major faults. Folding of rocks can assist flow of fluids, either because of associated fracturing or because movement on the folded beds can give rise to openings. You can see how this works by arching a thick stack of writing paper. Openings form between the sheets at the crest of the 'anticline'. If orebodies form in these openings they are called *saddle reefs*. Further structurally favourable environments are breccia zones. These are caused by a variety of processes, including: high-velocity upward movement of gaseous fluids known as *fluidization;* caving and collapse due to magma withdrawal below; collapse due to the dissolving of underlying formations; and crushing and shearing due to tectonic movements. Mineralized *breccia pipes*, veins, dykes

saddle reefs

fluidization

breccia pipes

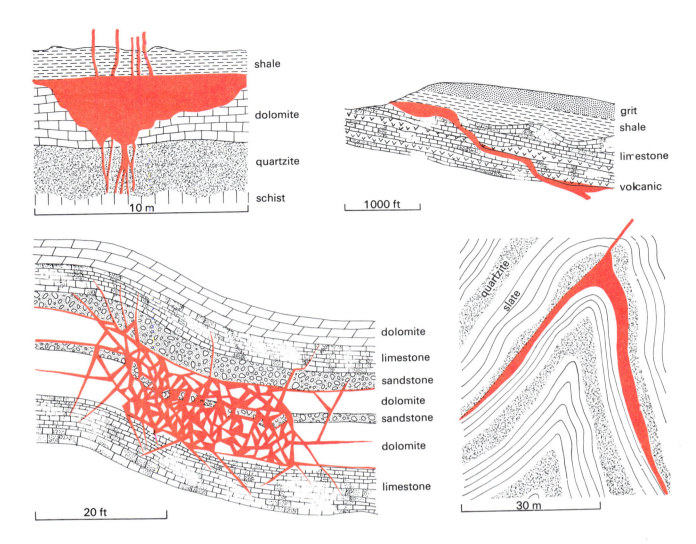

*Figure 37* Orebodies showing structural controls on ore deposition (the ore is marked in red) (for use with ITQ 33).

and channels are often the result. Finally, it has been suggested that ore fluids can migrate by diffusion of water molecules through solid rock, although it is difficult to imagine how very large volumes of fluid could move in this way.

ITQ 33   For each of the orebodies shown in Figure 37 a–d identify the main structural controls on ore deposition.

Now we can examine the chemical controls on ore deposition. You should be aware from reading about hydrothermal solutions in S26–, Block 3, Section 3.5.2, that metals are carried in solution as complexes with ions such as $Cl^-$, $SO_4^{2-}$, $HS^-$. Why, then, should metal sulphides suddenly precipitate from such solutions to form a mineral deposit? In S26–, we gave you one explanation—sudden contact between the solution and a source of reduced sulphur. However, hydrothermal mineral deposits can be found in such a wide range of geological environments that this is clearly not the only explanation.

A given metal complex in solution is only stable over a certain range of fluid composition, pH, temperature and pressure. Ore may precipitate if one of these parameters is changed. You have already come across temperature changes in connection with wall-rock alteration. The three methods of cooling hydrothermal solutions were given then as: loss of heat to wall rock; mixing with cool circulating groundwaters; and expansion. A drop in temperature will reduce the stability of the complex ion, until ultimately the metal sulphide is precipitated. Pressure changes can take place simply because the fluid is rising to shallower levels in the crust, or they can take place quickly if the fluids are able to expand suddenly—as they might in a wide fracture or breccia zone for instance. One way that this could lead to ore deposition is by releasing gases such as $H_2S$ or $CO_2$ from solution, thus allowing them to react with metal ions; compared with temperature and chemical changes, however, this is probably of minor importance.

**chemical controls on ore deposition**

There are many chemical changes capable of causing ore deposition. Exactly which of these are the most important ones is a matter of some controversy and depends largely on the nature of the complexes that are keeping the metals in solution. The use of computers to make thermodynamic calculations has led to considerable advances in this field for they enable the behaviour of these complexes in hydrothermal fluids to be predicted. Here we shall review very briefly the current ideas on the subject.

In many deposits, and El Salvador was probably one, the metals were kept in solution as chloride complexes. These complexes are most stable when the fluid is very acid. If some reaction takes place which increases the pH of the fluid, the ultimate result is to reduce the stability of the complex and release metal ions. Given a source of sulphur in the fluid, metal sulphides will be precipitated. In other words, a reaction such as:

$$CuCl_3^- \rightleftharpoons Cu_2^+ + 3Cl^-$$

will tend to move to the left in fluids of low pH but move to the right as pH increases.

Can you recall from Section 5.3 a reaction that could cause such a pH increase?

Wall-rock alteration of potash feldspar to sericite (equation 1) or hornblende to biotite (equation 2) are two good examples. Other possibilities might include the reaction of fluids with a limestone wall rock or the mixing of a low-pH fluid with one of higher pH. (Equations 1 and 2 are on p. 46.)

However, in some deposits, other complexing species such as polysulphide or carbonate may be of the greatest importance. If polysulphide complexes are keeping metals in solution, then *lowering* of pH or oxidation may be the critical factors in promoting ore deposition. Lowering of pH means the increasing of hydrogen ion concentration in the fluid, and this can then take part in the following reactions:

$$H^+ + S^{2-} \rightleftharpoons HS^-$$
$$H^+ + HS^- \rightleftharpoons H_2S_{gas}$$

This reduces the concentrations of the complex-forming species, $HS^-$ and $S^{2-}$. Oxidation leads to the generation of hydrogen ions so that this process can operate. If we denote an oxidizing agent by $O_2$, we get:

$$H_2S + 2O_2 \rightleftharpoons SO_4^{2-} + 2H^+$$
$$HS^- + 2O_2 \rightleftharpoons SO_4^{2-} + H^+$$
$$S^{2-} + 2O_2 \rightleftharpoons SO_4^{2-}$$

So why did ore minerals precipitate at El Salvador?

ITQ 34   Which of the following factors do you think might have been effective in causing (a) Early and (b) Late mineralization at El Salvador?

1   decrease in temperature

2   decrease in pressure

3   mixing of magmatic and meteoric water

4   reaction with wall rock

## 7.3   A genetic model for El Salvador

You are now in a position to synthesize the El Salvador data to produce a genetic model for the deposit. You can do this using the following headings:

1   *Form and setting of the deposit*   Describe the geological setting in terms of geological environment, structures and rock types. Describe the form of the deposit, i.e., the way that the ore minerals are distributed within the host rock.

genetic models for ore deposition

2   *Zoning*   Describe the evolution in time and space of ore minerals within the deposit.

3   *Wall-rock alteration*   Record the types of wall-rock alteration and their distribution (zoning) within the ore deposit.

4 *Nature and origin of the fluids* Record the results of fluid inclusion and isotopic studies related to the possible source or sources of the fluids and their salinities.

5 *Origin of the ore constituents* Record any evidence for the origin of the ore constituents, usually metals and sulphur.

6 *Temperature of the fluids* Give the possible range of temperatures for the mineralizing fluids. These data will most probably have been derived from fluid inclusion or isotope studies, or detailed studies of the mineral phases present.

7 *Causes of ore deposition* Explain the physical and chemical controls operating during the formation of the ore deposit. Explain also the phenomena of mineral zoning and wall-rock alteration.

8 *History of the ore deposits* Briefly summarize the history of the deposit, using diagrams where necessary.

Of course, El Salvador has had a complicated history involving several phases of hypogene mineralization followed by supergene enrichment. You will therefore need to describe *each* phase of mineralization in terms of headings 2–7.

ITQ 35 By completing Table 15, summarize the genesis of the El Salvador porphyry copper deposit. You can use this exercise for revision of the earlier parts of the text.

TABLE 15 A summary of the genesis of the El Salvador porphyry copper deposit (to be completed as ITQ 35).

| | Early mineralization | Transitional mineralization | Late mineralization | Supergene enrichment |
|---|---|---|---|---|
| Form of ore | | | | |
| Age | | | | |
| Ore minerals and zoning | | | | |
| Alteration | | | | |
| Source and nature of fluids | | | | |
| Source of metals and sulphur | | | | |
| Temperature | | | | |
| Controls on ore deposition: chemical physical | | | | |

So we can describe the history of the El Salvador deposit as follows:

The hypogene mineralization at El Salvador was associated with the intrusion of several porphyritic granodiorite stocks, about 41 Ma ago. The Early mineralization was associated with the first two porphyries (the *X*- and *K*-porphyries) which produced the *A* quartz veins, disseminated sulphide mineralization and K-silicate alteration. Both mineralization and alteration exhibited zoning: the sulphide mineralization varies from chalcopyrite–bornite at the centre to chalcopyrite–pyrite at the periphery; the K-silicate alteration zone was surrounded by a zone of propylitic alteration (Fig. 38). Fluid inclusions suggest

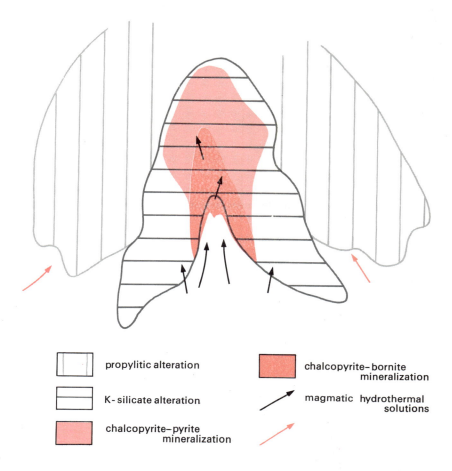

| | | |
|---|---|---|
| propylitic alteration | | chalcopyrite–bornite mineralization |
| K-silicate alteration | | magmatic hydrothermal solutions |
| chalcopyrite–pyrite mineralization | | |

*Figure 38* Early alteration and mineralization at El Salvador (before intrusion of the *L*-porphyry) (from Gustafson and Hunt, 1975).

that early mineralization was carried out by hot (350–600 °C) saline fluids, and isotopic studies point to a magmatic origin for much of this water. The majority of the copper, and perhaps also the sulphur, was probably also derived from the magma, but we cannot tell whether the ultimate source was deep continental crust, mantle or subducted oceanic crust. Deposition of the ore minerals could have resulted from changes in fluid chemistry due to wall-rock alteration and decrease in temperature and pressure; the alteration mainly involved $K^+$ ions in the fluids replacing $Ca^{2+}$ and $Na^+$ ions in the wall rock. Much of the sulphur was fixed as anhydrite. The structural controls for fluid migration were provided by stockwork fractures caused by hydraulic fracturing of the country rock, and by inter-grain permeability.

The final porphyry magma, *L*-porphyry, followed most of the early stage mineralization. Transitional mineralization followed, carrying most of the molybdenum, which was deposited in *B* quartz veins. Mineralization was again by hot (350–600 °C) saline fluids, but with a somewhat greater component of meteoric water. Much of the structural control was provided by flat-dipping fractures, perhaps related to cooling and shrinking of the intrusive.

Late mineralization followed the Transitional mineralization and was dominated by pyrite, both disseminated (in a pyrite and a pyrite–bornite zone) and in *D*-veins. Accompanying alteration was dominantly sericitic (Fig. 39). Fluid inclusions suggested that the fluids were of low temperature (< 350 °C) and low salinity. Isotopic data suggested that meteoric water was a major component of the mineralizing fluids; this had worked its way downwards through fractures, remobilizing early sulphides and anhydrite and redepositing them as **Late** mineralization.

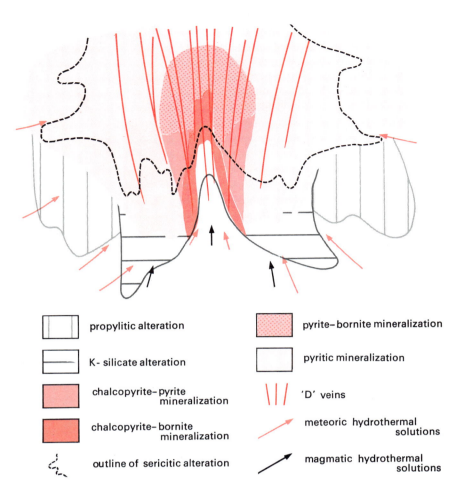

| | |
|---|---|
| propylitic alteration | pyrite–bornite mineralization |
| K-silicate alteration | pyritic mineralization |
| chalcopyrite–pyrite mineralization | 'D' veins |
| chalcopyrite–bornite mineralization | meteoric hydrothermal solutions |
| outline of sericitic alteration | magmatic hydrothermal solutions |

*Figure 39* Main period of late alteration and mineralization (after intrusion of the *L*-porphyry) (from Gustafson and Hunt. 1975).

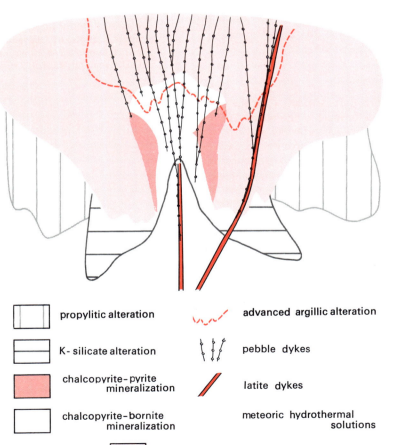

| | |
|---|---|
| propylitic alteration | advanced argillic alteration |
| K-silicate alteration | pebble dykes |
| chalcopyrite–pyrite mineralization | latite dykes |
| chalcopyrite–bornite mineralization | meteoric hydrothermal solutions |
| pyritic mineralization | |

*Figure 40* Very late post-mineralization hot spring state (during intrusion of latite) (from Gustafson and Hunt, 1975).

As cooling of the intrusive centre continued, meteoric waters took over completely and caused advanced argillic alteration and hot-spring activity. A final and minor advance of magma formed pebble dykes, and latite dykes (Fig. 40).

Supergene enrichment of the hypogene copper ores followed this hot-spring stage of primary mineralization. Metals leached from the surface rocks (producing a leached capping) percolated downwards in solution until they reached the water table. *In situ* replacement of chalcopyrite and bornite to secondary Cu–S minerals such as chalcocite gave rise to the commercial orebody. These highly acid waters produced a zone of argillic alteration, selectively destroying feldspars, biotite and chlorite in the zone of enrichment. Erosion and oxidation of the supergene orebody followed and continues today.

## 8.0  Porphyry copper exploration

**Study comment**  Anaconda invested extra manpower and funds in the detailed study of El Salvador in order to improve our understanding of the processes of ore deposition and to develop some new tools and concepts for exploration for porphyry copper deposits both in Chile and elsewhere in the world. In this Section we shall summarize the implications of the El Salvador study for porphyry copper exploration. We shall then look briefly at other porphyry copper deposits, including one example from Britain, and compare their similarities and differences, particularly as they relate to the search for new deposits.

### 8.1  Guides to ore search

In Section 2, on porphyry copper resources, we explained that there is now a great challenge to Earth scientists to apply geological reasoning to predict locations of mineral resources concealed in the sub-surface. Let us, therefore, consider briefly the types of regional and local guides to ore that the mining companies must look for.

The *regional* guides give the mining companies ideas on where to send their exploration teams. On the broadest scale, these teams must identify the correct geological environment for the type of ore they are looking for. Within this area they must look for favourable rock types or structures. A company may also carry out a literature search and field reconnaissance to discover whether any old mine workings exist in the area and to examine the distribution of mines that are being worked—are they associated with major faults or with rocks of one particular age, for instance? You should be familiar with these different lines of approach from S26–, Block 3, Sections 5 and 6.

It is now the job of the exploration team to look for some of the *local* guides to ore. McKinstry (1948) classified local guides under four headings: *physiographic guides; mineralogical guides; stratigraphic and lithologic guides; structural guides.*

*Physiographic guides* do not only include topographic features that result directly from the presence of orebodies, although these do exist (oxide copper deposits in the Atacama desert of Chile often stand out as small bluish hills, for instance). The term physiographic also covers topographic features which are favourable to ore formation. This has particular relevance to secondary enrichment.

> Which of the following favour(s) secondary enrichment: a deep water table; a shallow water table; a rapid·rate of erosion; a slow rate of erosion?

Clearly, a deep water table and a slow rate of erosion will favour secondary enrichment. In the latter case, a high rate of physical removal of rock material will not give time for the necessary chemical reactions to take place. And a shallow water table not only means that there is less rock to be leached by the groundwater, but also implies a good deal of lateral movement of groundwater into stream valleys with loss of dissolved elements. The topography, which is partly related to rainfall, water circulation and erosion rate, will therefore have a strong bearing on the extent of secondarily enriched ores.

The best *mineralogical guides* are, of course, the ore minerals themselves. However, from what you have read about El Salvador, you should be able to suggest several other valuable guides that may extend over a wider area than the actual ore deposit.

> From your understanding of this and other courses (especially S26–), list three mineralogical guides to ore that you think might be important.

guides to ore

physiographic guides

mineralogical guides

*Leached and oxidized cappings*, resulting from the interaction of groundwater with the orebody are particularly important. Several books have been written on the study of iron-rich leached cappings (or gossans) as a way of predicting the nature of the underlying rocks. *Wall-rock alteration* is another obvious feature, commonly represented by complete changes in mineralogy, though in some rocks—especially relatively pure sandstones and limestones—the changes may be solely in texture and colour. The zoning of alteration mineral assemblages is the most useful guide to ore—rather than the simple knowledge that altered rocks exist. *Related primary mineralization* such as pyritic zones and gangue minerals are a third guide to the presence of minerals of economic value.

*Stratigraphic and lithologic guides* refer to rock formations and rock types that are favourable for ore deposition. You have already come across some of these in Section 7.2. Good examples of particularly favourable rock types are: breccias and conglomerates; limestones; black shales.

**stratigraphic guides**

> What property characterizes each of these three rock types to make them likely hosts for ore deposits?

Breccias and conglomerates are permeable rocks through and into which mineralizing fluids can readily migrate; limestones are chemically reactive rocks, which can be dissolved by ore fluids and replaced by ore minerals; black shales carry reduced sulphur and can thus induce precipitation of ore minerals from hydrothermal solutions.

*Structural guides* include fractures, geological contacts, and folds. These structures are all capable of controlling the flow of mineralizing fluids and can therefore be favourable loci for ore deposition in specific cases. In the case of some vein deposits, certain fractures will be preferentially mineralized with respect to other fractures of different orientation, so the analysis of fracture patterns can be useful.

**structural guides**

In the next Section, we shall concentrate on those features that can be applied specifically to porphyry copper exploration. Many of these features were well illustrated in the earlier Sections on El Salvador.

## 8.2 Evidence from El Salvador

What, then, are the implications of the El Salvador study for porphyry copper exploration? Dr John P. Hunt, co-author of the original paper on the El Salvador deposit, former Chief Geologist of The Anaconda Company and Vice President of Anaconda's Primary Metals Division, has provided an answer to this question for use in this *Case Study*. We have reproduced his comments below:

> The details of the anatomy and evolution of the El Salvador deposit can be applied in exploration for porphyry copper deposits. They can be applied primarily in cases where some portion of the gross body of mineralization associated with a porphyry copper is exposed at the surface or in a prospect drill hole or underground working. The El Salvador studies can *not* readily be applied to the search for completely concealed, or 'blind' deposits. The search for blind deposits is severely restricted at present by a fundamental lack of knowledge of the relationship between ore formation and broad regional and crustal processes. As illustrated in a later Section of this *Case Study*, one is only capable of vaguely identifying whether or not a broad region is a potential porphyry copper terrain. We cannot predict specifically where to look within that terrain unless mineralization is exposed at the surface. Nor can we predict with assurance whether an unproven porphyry copper province will contain large or small deposits, high grade or low grade, and so on. Thus there is much room for imaginative research and new ideas to close the gap between detailed work of the El Salvador type and broad, regional studies and theory, such as modern plate tectonics.

> The evidence from El Salvador which is of most direct value in exploration can be divided into three general categories: (1) zonal patterns of sulphide mineralization and wall rock alteration; (2) rock type and structural features indicative of centres of mineralization; and (3) mineralogical and chemical data from the leached capping indicative of supergene enrichment.

### 1 Zonal patterns of mineralization and alteration

Studies of the El Salvador orebody have shown that primary alteration assemblages of K-feldspar–biotite, sericite–chlorite, sericite–andalusite tend to occur preferentially but not invariably with the following primary sulphide assemblages respectively; chalcopyrite–bornite, pyrite–chalcopyrite, and pyrite–bornite–chalcopyrite. Argillic (kaolinitic) alteration is secondary and associated with supergene replacement of primary copper–iron sulphides by Cu–S minerals such as chalcocite–djurleite or djurleite–digenite. In general, primary sulphide assemblages interpreted as reflecting higher sulphur activity are associated with alteration assemblages representing conditions of higher ratios of hydrogen ion to alkali (K, Na) ion activity. The zonal patterns of both sulphides and alteration products point to increasing sulphur concentrations and higher $H^+/(K^+, Na^+)$ ratios upwards and outwards from the zonal centre of the deposit.

These generalizations provide a rough three-dimensional framework for interpreting alteration patterns encountered in exploration drill holes or surface mapping. Thus, for example, when pyritic mineralization and strong sericite alteration are encountered, one may generally assume that a penetration of the upper or peripheral portions of a deposit has occurred. A shift in drill sites can then be planned to penetrate an inner and higher grade zone of mineralization, with a minimum number of drill holes.

### 2 Rock type and structural patterns

A number of intimately associated porphyry intrusions, with K-silicate alteration and low-sulphur chalcopyrite–bornite primary mineralization, mark the central or core zone of the El Salvador deposit. In exploration, multiplicity of porphyry rock types is in itself a favourable indicator. The zonal centre of the deposit may be better defined by a point about which porphyry bodies are clustered than by the outline of any one intrusive. Other favourable structural features which indicate the more favourable zone of mineralization are: zones of 'A' quartz veins; zones of intense replacement of plagioclase by a perthitic, K-feldspar groundmass; and bodies of igneous breccias. In exploration, drill hole target selection must consider these and other features as well as the patterns of mineralization and alteration, if a given prospect is to be adequately and efficiently tested.

### 3 Limonite and chemical data

Interpretation of leached cappings has long been more of an art than a science. The chemical patterns present in the outcrop and the nature of the limonite resulting from oxidation of sulphide minerals is a composite effect of physiography, climate and several variables in the rocks themselves, such as the solubility of the various metals, the acid-generating capacity of the mineralization, and the neutralizing capacity of the wall rock. Thus, each capping is essentially a unique composite product of these many variables.

At El Salvador, because of the strong zoning of the primary mineralization, a variety of leached capping types was found to overlie secondarily enriched ore. Thus, depending on position with respect to primary zoning, and other indicators such as 'A' quartz veins and porphyry rock types, it is possible to classify several markedly different types of cappings as 'favourable', At El Salvador, the majority of the leached capping is within the zone of sericitic alteration (with the major exception of the portion occupied by the L-porphyry intrusive). The sericitic rocks were essentially 'neutral' to acid supergene waters. Original variations in pyrite content are preserved as differences in total iron content and limonite abundance in the capping but molybdenum and gold are relatively immobile. Thus a general pattern is present, like a giant bulls-eye, with an area of higher iron and limonite on the outside, surrounding a central area with lower iron that contains relatively high Mo and Au values, and overlies the thickest portions of the enrichment blanket.

## 8.3 Other porphyry copper deposits

Of course, El Salvador is only one of more than a hundred known porphyry copper deposits (see Fig. 1), and it would be dangerous indeed to base an exploration programme on the knowledge gained from just one deposit. As Gustafson and Hunt themselves warned in their paper: 'We urge caution to those who apply a "typical" porphyry copper model to the solution of major geological problems and especially to those who are responsible for wisely investing their company's exploration funds!' So let us just briefly examine the sort of variations that the exploration geologist might come across. On the following pages we summarize five descriptions of porphyry copper deposits that have appeared in the geological literature.

variations between porphyry copper deposits

65

### 1 Chuquicamata (after Perry, 1952, and Sillitoe, 1973)

Chuquicamata in Northern Chile (Fig. 41) is one of the world's largest copper mines. It is related to a series of Tertiary granodiorite porphyries which have invaded volcanic and sedimentary strata. The hypogene mineralization consists of abundant pyrite and enargite with subordinate chalcopyrite, chalcocite and molybdenite. This mineralization occurs within granodiorite which has been affected by sericitic alteration and silicification; this appears to have obliterated an earlier K-silicate alteration event. The sericitic zone grades eastwards into propylitic alteration and finally into fresh granodiorite. It is cut off sharply westwards by the West Fissure, the crush zone of a major fault which has removed part of the orebody and brought fresh granodiorite to the surface. The orebody itself is a thick oxidized layer of copper sulphates underlain by a chalcocite-bearing enrichment blanket.

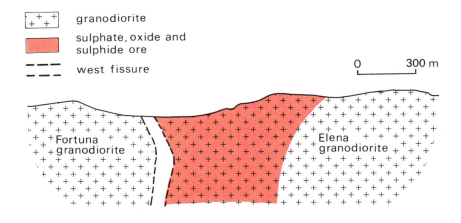

granodiorite

sulphate, oxide and
sulphide ore

west fissure

0          300 m

*Figure 41* A section through the Chuquicamata orebody (after Perry, 1953, and Sillitoe, 1973).

### 2 El Teniente (after Howell and Molloy, 1960)

The El Teniente mine in central Chile (Fig. 42) has been exploited since about 1908 and is the largest underground copper mine in the world. The orebody is related to Tertiary intrusions of quartz diorite and dacite porphyry which cut a thick pile of andesites and terrestrial sediments. The main structural feature in the mine area is a breccia pipe known as the Braden Pipe, an inverted cone-shaped cavity in the volcanic rocks filled with rock debris. The deposit itself surrounds this pipe, which appears to have post-dated and destroyed much of the mineralization and alteration. The most obvious wall-rock alteration now visible is abundant secondary biotite throughout the andesite within the orebody. A number of phases of primary mineralization can be recognized, the main phase involving deposition of chalcopyrite and bornite with quartz–anhydrite gangue. Subsequent secondary enrichment has increased the grade of the primary ore almost twofold.

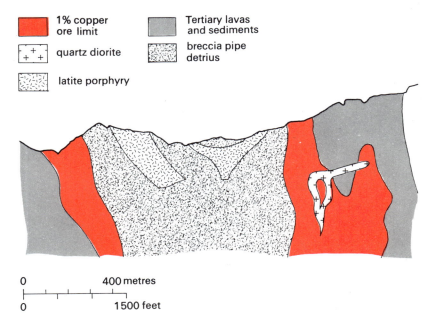

1% copper
ore limit

quartz diorite

latite porphyry

Tertiary lavas
and sediments

breccia pipe
detrius

0          400 metres

0          1500 feet

*Figure 42* A section through the El Teniente orebody (from Howell and Molloy, 1960).

## 3  *Los Pelambres* (after Sillitoe, 1973)

The Los Pelambres deposit in Central Chile (Fig. 43) is related to an Upper Miocene tonalite stock which was passively emplaced into Mesozoic volcanic and sedimentary strata. A zonal pattern of alteration and primary mineralization is centred on the stock. The core zone displays K-silicate alteration together with chalcopyrite, bornite, pyrite and molybdenite mineralization and quartz–anhydrite gangue. The marginal part of the stock has undergone sericitic and argillic alteration, with pyrite as the main sulphide mineral. Small bodies of igneous breccia can be found, mainly in the peripheral zones. Supergene enrichment is poorly developed at Los Pelambres because of its relatively recent age of formation and the effect of intense glacial erosion.

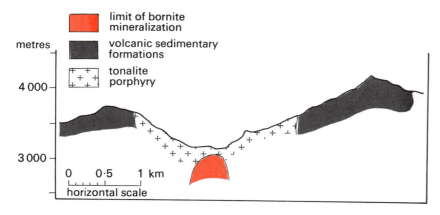

*Figure 43*  A section through the Los Pelambres orebody (from Sillitoe, 1973).

## 4  *Mount Fublian* (after Bamford, 1972)

Mount Fublian (Ok Tedi) (Fig. 44) is an exceptionally young porphyry copper deposit recently discovered on the island of New Guinea. It is related to small Pliocene/Pleistocene quartz diorite and granodiorite stocks which intrude folded sedimentary strata. The central mineralized intrusion exhibits intense K-silicate alteration. The dominant primary copper sulphide is chalcopyrite, associated with bornite, marcasite (a form of $FeS_2$), pyrite, significant gold and silver, and minor molybdenite.

No halo of sericitic alteration has been recognized. Secondary enrichment has produced a thick chalcocite enrichment blanket beneath a leached and oxidized capping. Closely associated with this disseminated mineralization are small bodies of massive magnetite and sulphides within limestone country rocks.

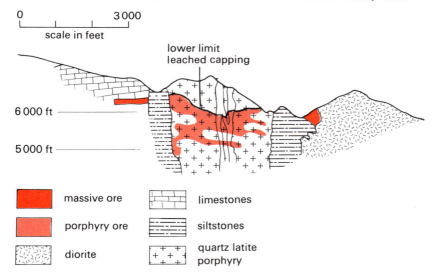

*Figure 44*  A section through the Mt Fublian orebody (from Bamford, 1972).

## 5  *San Manuel–Kalamazoo* (after Lowell and Guilbert, 1970)

The San Manuel–Kalamazoo deposit in Arizona (Fig. 45) is centred around Early Tertiary (65 Ma) monzonite porphyry and dacite porphyry stocks, which have intruded into Cretaceous sediments and Precambrian granite. The mineralization occurs equally within the Tertiary and Precambrian granitic rocks. Concentric zones of alteration and mineralization are centred on the monzonite porphyry. Wall-rock alteration ranges from K-silicate type at the centre through sericitic and argillic to propylitic at the periphery. Primary copper mineralization, which is restricted to the K-silicate and sericitic zones, comprises chalcopyrite with pyrite and molybdenite and forms a shell around a low-grade core. Pyrite

67

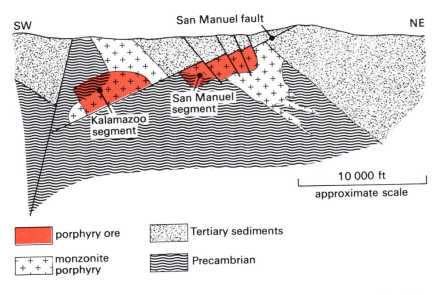

SW                    San Manuel fault                    NE

San Manuel fault

Kalamazoo
segment

San Manuel
segment

10 000 ft
approximate scale

porphyry ore          Tertiary sediments

monzonite
porphyry              Precambrian

*Figure 45*  A section through the San
Manuel–Kalamazoo orebody (from
Lowell and Guilbert, 1970).

is the dominant sulphide around the periphery of the deposit. The deposit has
suffered several post-mineralization events: first tilting, then some secondary
enrichment, more tilting and finally displacement by the San Manuel fault. It was
this fault that divided the deposit into two roughly equal portions, the San Manuel
and the Kalamazoo orebodies.

ITQ 36  These five deposits each exhibit at least one major geological difference
with respect to El Salvador. Which of these deposits are distinguished from El
Salvador by the following features:

1  Poorly developed secondary enrichment?

2  Poorly developed sericitic alteration?

3  Poorly developed K-silicate alteration?

4  Associated contact metasomatic copper orebodies?

5  Loss of part of the orebody through a late volcanic event?

6  Distortion of mineralization and alteration patterns through post-mineralization tectonic event?

According to Gustafson and Hunt, differences such as these are only 'variations
on a theme'. The same processes, and evolutionary trends can be observed for
each deposit, even though variations in the details of the genetic model or in
post-mineralization events may have given it a unique geological character.
Gustafson and Hunt listed three all-important processes for development of
porphyry copper deposits:

1  Relatively shallow emplacement of a series of small porphyritic intrusions
related to an underlying batholith.

2  Introduction of mineralizing fluids from the solidifying magma into both
porphyries and country rock.

3  Interaction of circulating groundwater with the cooling mineralized centre.

Some of the differences between the various porphyry copper deposits can be
explained in terms of slight variations in these processes. ITQ 37 illustrates one
way in which this can happen.

ITQ 37  Consider a deposit where the mineralizing porphyry crystallized at great
depth.

(a) How will this affect the interaction of groundwater with the mineralized
centre?

(b) How will this affect the nature and extent of sericitic alteration and pyrite
mineralization around the deposit?

Other differences can be explained in terms of post-mineralization events. You
encountered such events as faulting and late magmatic activity in the five
examples given earlier. However, erosion may be the single, most important

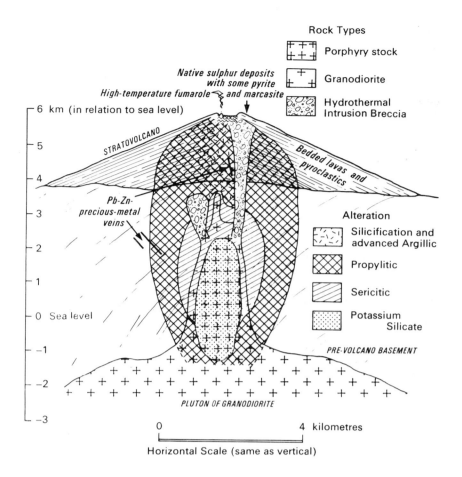

**Rock Types**

⊞ Porphyry stock

▭ Granodiorite

▨ Hydrothermal Intrusion Breccia

Native sulphur deposits with some pyrite and marcasite

High-temperature fumarole

6 km (in relation to sea level)

STRATOVOLCANO

Bedded lavas and pyroclastics

Pb-Zn-precious-metal veins

**Alteration**

▨ Silicification and advanced Argillic

▨ Propylitic

▨ Sericitic

▨ Potassium Silicate

PRE-VOLCANO BASEMENT

PLUTON OF GRANODIORITE

0 ———— 4 kilometres
Horizontal Scale (same as vertical)

*Figure 46* Sillitoe's model showing the effect of erosion on a porphyry copper deposit (from Sillitoe, 1973).

factor. Sillitoe (1973) pointed out that depth of erosion can determine what part of a mineralized column is preserved. His model is given in Figure 46.

**ITQ 38** According to Sillitoe's model would you expect to find extensive sericitic alteration exposed by shallow or deep levels or erosion?

So the geologist engaged in exploration for porphyry copper deposits must keep in mind all these possible deviations from the idealized porphyry copper model. As we emphasized in Section 5, however, it is more important for you to understand the processes involved than to recall the precise details from any one deposit.

## 8.4 Porphyry copper deposits in Britain

This final Section is designed to set you thinking about the possibilities of finding porphyry copper deposits in the British Isles. This subject raises a number of economic and environmental questions, and we dealt with some of these in S26–, *The Earth's Physical Resources*. In this Section we shall concentrate on the geological aspects. The first step is to look at the regional picture, and so narrow down the area worth exploring by locating regions where favourable rocks outcrop—that means regions with abundant granitic rocks.

*Try to locate such areas on your geological map of the British Isles* (included with S23–, Block 6, *Historical Geology*).

There are six main areas, marked in red on your map:
1. The British Tertiary Province
2. The Scottish Highlands
3. The Southern Uplands – E Ireland
4. The Lake District
5. North Wales–SE Ireland
6. SW England

69

These areas represent a variety of past tectonic environments, some undoubtedly more favourable for porphyry copper formation than others.

*By using S23–, Block 6, for reference, try to decide which of these regions were related to destructive plate boundaries and, therefore, might be potential porphyry copper provinces.*

As you now know from reading the *Igneous Case Study*, the *British Tertiary Province* was related to a constructive rather than destructive plate boundary and contains a relatively small proportion of granitic rocks. So this is the least likely place for finding porphyry copper deposits. Geologists are divided on the original tectonic setting of the *SW England province*. It seems unlikely to contain porphyry copper deposits, because it is a copper–tin rather than copper–molybdenum province, and because no such deposits have been discovered during the intensive exploration for vein deposits. We are therefore left with areas 2–5 for further consideration. Each of these was probably associated with a destructive plate boundary of Caledonian age (S23–, Block 6, Sections 3.2.2 to 3.2.4) and so the regional setting was right. The main question is whether these rocks have been too deeply eroded for any deposits to be preserved. Porphyry copper deposits of this age are rare, although a few are in production elsewhere in the world including the 400 Ma deposit at Gaspe in NE United States.

It is clear that some of the Caledonian granites have been deeply eroded, particularly those in the Scottish Highlands, but there are also many examples of shallow erosion where overlying volcanic rocks have also been preserved. Moreover, copper–molybdenum mineralization has been observed in a number of these granites. You may have visited the Shap granite and recalled that it contains disseminated chalcopyrite with molybdenite on joint faces; several of the

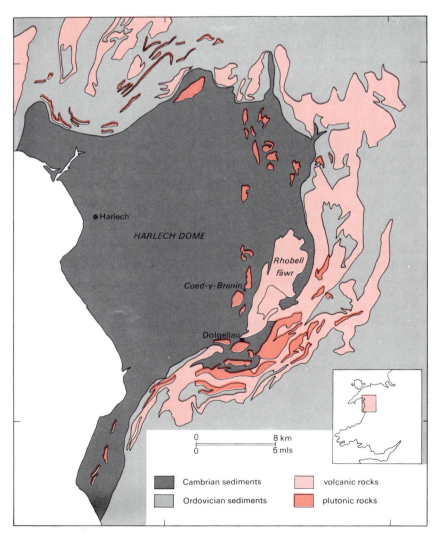

*Figure 47* Regional setting of the Coed-y-Brenin prospect, North Wales.

granites in the Scottish Highlands are also known to contain abnormally high concentrations of molybdenum. By far the most important discovery so far, however, has been the potential porphyry copper deposit in the forest of Coed-y-Brenin in North Wales, the result of an exploration programme conducted by Rio Tinto Zinc between 1968 and 1972. Let us now examine this, the only porphyry copper deposit so far discovered in Britain, in more detail.

The Coed-y-Brenin deposit lies on the eastern margin of the Harlech dome, a structure which has led to the exposure of sediments of Cambrian age. In the Coed-y-Brenin region, the Cambrian argillaceous sediments are intruded and in places partly assimilated by a series of small diorite intrusions some of which are porphyritic. It is thought that these diorites may be comagmatic with the andesites of Rhobell Fawr, a volcano of Upper Cambrian or Lower Ordovician age that lies to the west (Fig. 47).

The area has been mined for copper, gold, lead and zinc in the past and it was one of the old mines, Turf Copper, which provided the company with the first clue to the existence of a much larger orebody. Geochemical surveys followed by a drilling programme enabled it to be located and evaluated. A preliminary evaluation indicated about 200 million tons of 0.3 per cent copper ore with subordinate molybdenite and gold.

ITQ 39   Figure 48 covers part of the Coed-y-Brenin forest, showing the mineralization and alteration patterns observed at the surface. Using your knowledge from Sections 8.1 and 8.2, sketch on to this map the area where you think the orebody is likely to be.

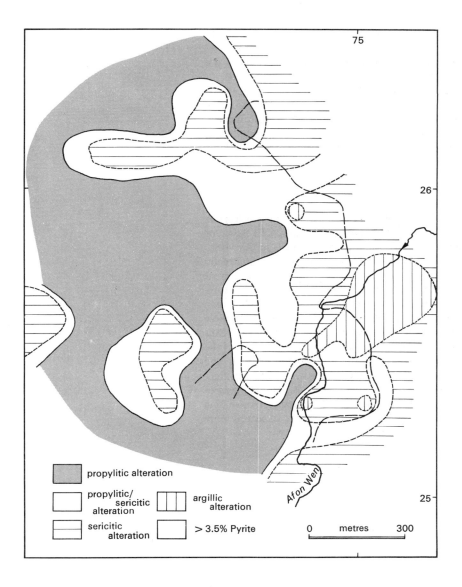

propylitic alteration

propylitic/sericitic alteration

sericitic alteration

argillic alteration

> 3.5% Pyrite

0   metres   300

*Figure 48*   Alteration pattern at the Coed-y-Brenin prospect (from Rice and Sharpe, 1976).

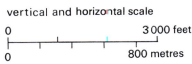

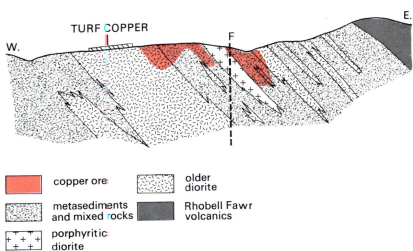

*Figure 49* A section through the Coed-y-Brenin prospect (from Rice and Sharpe, 1976).

The geology of this deposit has been very much obscured by later tectonic activity and metamorphism. However, you should have noticed from ITQ 39 that some of the main geological indicators have been preserved. Figure 49 shows a cross-section through the deposit, as constructed by the Rio Tinto-Zinc geologists.

What are the main implications of the discovery of the Coed-y-Brenin deposit?

It confirms that Caledonian igneous provinces in Britain, and particularly the province extending from North Wales into SE Ireland, may provide worthwhile targets for porphyry copper exploration. However, it may be some time before further deposits are found. One effect of the environmental controversy over Coed-y-Brenin was to cause a cutback in this type of exploration in Britain. Nevertheless, future British governments faced with rising costs of importing copper may be forced to reopen the debate over the environmental constraints on mining, and we may yet see a porphyry copper mine in the British Isles.

**Concluding study comment**

Now that you have read this *Case Study*, you should be aware that – at least as far as ore deposits are concerned – the Earth is not a simple place. El Salvador is one of the best exposed and most thoroughly studied ore deposits in the world. Yet its discovery was certainly not without its problems; and there are still many aspects of its genesis that are not well understood. There is, therefore, great scope in economic geology for research of both 'academic' and 'applied' nature, and at Summer School you may have the opportunity of applying your own imagination to some of these unsolved problems. However, in spite of these constraints, a considerable understanding of a deposit can be built up by means of the approach presented in this *Case Study*: first, a detailed examination of the field relations; second, microscopic study of the ore and gangue minerals and of the host rock; third, laboratory studies of these minerals, particularly their fluid inclusions and isotopic composition; and finally analysis and synthesis of these data in order to construct a model for its origin. And, hopefully, by reaching an understanding of how mineral deposits form we shall be able to search more effectively for new deposits.

# References

Bamford, R. W. (1972) The Mount Fublian (Ok Tedi) porphyry copper deposit, Territory of Papua and New Guinea. *Econ. Geol,* **67,** 1019–1033.

Committee on mineral resources and the environment (COMRATE) (1975). *Mineral resources and the environment.* National Academy of Sciences, Washington.

Flawn, P. T. (1966) *Mineral Resources.* John Wiley.

Gustafson, L. B. and Hunt, J. P. (1975) The porphyry copper deposit at El Salvador, Chile. *Econ. Geol.* **70,** 859–912.

Howell, F. H. and Molloy, J. S. (1960) Geology of the Braden orebody, Chile, South America. *Econ. Geol.* **55,** 863–905.

James, D. E. (1971) Plate tectonic model for the evolution of the Central Andes. *Bull. geol. Soc. Am.,* **82,** 3325–3346.

James, D. E. (1973) The evolution of the Andes. *Scient. Amer.,* **229,** 60.

Lowell, J. D. and Guilbert, J. M. (1970) Lateral and vertical alteration-mineralization zoning in porphyry ore deposits. *Econ. Geol.,* **65,** 373–408.

McKinstry, H. E. (1948) *Mining Geology.* Prentice-Hall.

Parsons, A. B. (1933) *The Porphyry Coppers,* American institute of mining and metallurgical engineers.

Perry, V. D. (1952) Geology of the Chuquicamata orebody. *Min. Engng,* **4,** 1166–1169.

Perry, V. D. (1960) History of El Salvador development. *Min. Engng,* **12,** 339–343.

Rice, R. and Sharpe, G. (1976). Copper mineralization in the forest of Coed-y-Brenin, North Wales. *Trans. Instn Min. Metall.,* **85,** B1–B14.

Roedder, E. (1962) Ancient fluids in crystals. *Scient. Amer.,* **207,** 38–47.

Robbins, H. E., Dunstan, W. H., Dudley, T. H. and Pollish, L. (1960) Development of El Salvador mine. *Min. Engng,* **12,** 350–355.

Rose, A. W. (1970) Zonal relations of wallrock alteration and sulphide distribution at porphyry copper deposits. *Econ. Geol.,* **65,** 926–36.

Sillitoe, R. H. (1972) A plate tectonic model for the origin of porphyry copper deposits. *Econ. Geol.,* **67,** 184–197.

Sillitoe, R. H. (1973) The tops and bottoms of porphyry copper deposits. *Econ. Geol.,* **68,** 1–10.

Sillitoe, R. H. (1974) Tectonic segmentation of the Andes: implications for magmatism and metallogeny. *Nature, London.,* **250,** 542–545.

Swayne, W. H. and Trask, F. (1960) Geology of El Salvador. *Min. Engng,* **12,** 344–348.

## SAQ answers and comments

### SAQ 1

| | A<br>type of<br>deposit | B<br>usual host<br>rock | C<br>form of<br>the ore | Section in<br>*Mineral<br>Deposits* |
|---|---|---|---|---|
| 1 | Porphyry<br>Cu | (d) | (iii) and (iv) | 3.5.1 |
| 2 | Contact<br>metasomatic<br>Cu | (c) | all forms possible;<br>mainly (ii) | 3.3 |
| 3 | Vein Cu | any fractured<br>rock | (i) | 3.5.1 |
| 4 | Stratiform<br>Cu | (c) and (f) | usually (v) | 4.2.3 |
| 5 | Magmatic<br>segregation<br>Cu | (a) | usually (iv) and (v) | 3.2 |

6 Massive Cu-bearing sulphide deposits are associated with submarine volcanic rocks and are related to underwater volcanic activity. The ores grade from massive ore to a network of tiny veins or stockwork. Pyrite ($FeS_2$) is the predominant sulphide mineral with accessory copper (and often zinc and lead) sulphides.

7 Red-bed copper deposits are associated with intracontinental sediments, often fluviatile sandstones. The copper (and often uranium and vanadium) minerals are found infilling pores and veins in the host rocks, often replacing organic matter.

**SAQ 2** 1 C. This deposit is part of the Alpide belt and is associated with an apparent destructive plate margin of Tertiary age.

2 A. The Philippines are surrounded by oceanic crust and therefore represent a still active island arc, although the Atlas deposit itself is 60 Ma old.

3 B. This deposit is of Tertiary age and part of the still active Andean margin.

4 C. This part of the Rockies may have been the site of a Jurassic destructive plate margin.

5 C. The Bingham deposit, which is of Tertiary age, is one of a large number of deposits in western USA. Its original tectonic environment is obscured in part by faulting and volcanism that occurred after mineralization. Although you may well have chosen B for deposits 4 and 5, in fact the destructive plate margins in western North America are no longer active.

6 A. At 4 Ma old, Panguna is the youngest of the six deposits listed here and is in an active island arc setting.

**SAQ 3** 1 and 2: No. They tell us only about the distribution and nature of the ore minerals.

3 Yes. Fluid inclusions may be actual samples of the fluid from which ore minerals were deposited, which have been trapped during crystal growth.

4 Yes. The distribution of ore minerals within veins suggests deposition from fluids that have penetrated cracks in the host rock.

5 No. The presence of phenocrysts suggests slow cooling. Although this does also imply that the magma was hydrous, it is not sufficient evidence for a hydrothermal origin for the ores.

6 Yes. This suggests that the ores are related to fluids moving outwards from the igneous intrusion.

7 Yes. Hot aqueous fluids can cause extensive alteration of the rocks they pass through.

**SAQ 4** 1 Leached and oxidized zones. Limonite is among a number of iron-bearing minerals that form the leached and oxidized rock.

2 Primary zone.

3 Oxidized zone. Copper carbonates, phosphates and silicates such as malachite (green) and azurite (blue) impart a blue–green colour to the oxidized zone.

4 Leached zone. This is the colour of the residual hydrated iron oxides (limonite).

5 Enriched zone. The ratio of Cu to S in chalcocite is greater than in chalcopyrite.

6 Enriched zone. Occasionally, where an excess of copper has been introduced by solutions percolating downwards, the enrichment process can go all the way to cuprite ($Cu_2O$) or native copper (Cu).

**SAQ 5** 1 A. The lower grade, disseminated deposit is likely to require more crushing (*Mineral Deposits* (MD), Section 7.4.1).

2 B. Large-scale bulk mining methods are less effective for smaller, more concentrated ores (MD, Section 7.3.1).

3 A. More energy will be needed to extract, crush and concentrate the lower grade porphyry ores.

4 C. If the copper is in the form of sulphides in both types of ore, pollution will be a problem in both cases.

5 A. Because porphyry copper mining is carried out on a larger scale, a greater capital outlay is required.

6 B. A small sulphide-rich vein will give a steeper gradient than a large disseminated deposit (MD, Section 6.4.2).

**SAQ 6** 1 C (see Section 2.4.2).

2 A (see Table 4 and Section 2.4).

3 B or C. In Section 2.4 we stressed that, although the trend is towards very low-grade deposits, this may not continue into the future.

4 A (see Section 2.4.1).

5 C. There is the possibility that stratiform copper deposits will gain in importance and that manganese nodules mined by AD 2000 will produce a substantial proportion of the world's copper.

6 A or B (see Section 2.4.2). This will depend on our ability to apply geology to exploration.

**SAQ 7** Probably B, because the revolutionary change to large-scale mining methods took place over 70 years ago. C is also a valid answer, because current research into *in situ* mining and deep-sea mining may precede a further revolutionary step in mining technology.

**SAQ 8** Probably C, because mining companies are just beginning to think of ways of prospecting for 'blind' deposits and no consensus view is available. B is also a valid answer if you consider that plate tectonics (a revolution in a related field) is about to have a major impact on mineral exploration.

## ITQ answers and comments

### ITQ 1

| Class | average tonnage Cu (approx) | number of deposits | tonnage Cu contributed |
|---|---|---|---|
| 1 | $2 \times 10^7$ | 6 | $120 \times 10^6$ |
| 2 | $6 \times 10^6$ | 18 | $104 \times 10^6$ |
| 3 | $2 \times 10^6$ | 30 | $60 \times 10^6$ |
| 4 | $6 \times 10^5$ | 21 | $12.1 \times 10^6$ |
| 5 | $2 \times 10^5$ | 3 | $0.6 \times 10^6$ |
| 6 | $6 \times 10^4$ | 1 | $0.06 \times 10^6$ |

(a) Class 3 contains most deposits.

(b) Class 1 contains most copper.

(c) One class 1 deposit is probably more profitable (average tonnage $Cu = 2 \times 10^7$ compared with $8 \times 10^6$ for a class 2 and class 3 deposit). This emphasizes how important it is to a mining company to find these 'super-giant' deposits.

(d) You have met a similar diagram in consideration of Lasky's law (*Mineral Deposits*, Section 1.4). We shall be considering Lasky's law again in Section 2.3.

**ITQ 2** Table 3 shows cumulated production and reserves of $2.943 \times 10^9$ tons ore at a grade of 0.8 per cent Cu. According to Figure 2, this makes Bingham a class 1 deposit. Notice that the Bingham deposit was explored in steps; in contrast, some modern discoveries are evaluated thoroughly by drilling the entire deposit together with submarginal mineralization and this type of information, when available, greatly assists resource evaluation.

**ITQ 3** (a) The graph that you should have obtained (Fig. 50) approximates to a straight line and therefore fits Lasky's law very well.

(b) Assuming Lasky's law applies to lower grades, we get:
total copper reserves at 0.4 per cent $Cu = 3 \times 10^{10}$ tons ore
$= 1.2 \times 10^8$ tons Cu

This assumption may not, however, be valid. There is some evidence that the distribution of copper in a porphyry copper deposit is actually a bell-shaped curve with a peak at a grade of about 0.7 per cent Cu. This means that, although Lasky's law may hold for grades higher than 0.7 per cent Cu, it will not

hold at lower grades. So this prediction is probably over-optimistic and the real reserves are probably somewhat lower than this prediction would suggest.

(c) The lifetime of the deposit in years

$$= \frac{\text{copper left in the ground}}{\text{copper mined per year}}$$
$$= \frac{12.6 \times 10^6}{3 \times 10^5} = 42 \text{ years}$$

Of course, the actual lifetime may be much greater than this, and will depend in part on the grade of ore that can be mined in future and on the relationship between size of deposit and grade at ore grades less than 0.8 per cent Cu.

**ITQ 4** (a) Porphyry copper deposits form the majority of 'giant' and 'super-giant' deposits and these contain between them the great bulk of world copper reserves (see answer to ITQ 1). In this context, each new deposit of this type has a significant effect on world copper production, and there remain many relatively little explored parts of the world in which they could be discovered.

(b) In ITQ 3b we stressed that current data from the mining industry suggest that there is a 'threshold grade' of about 0.7 per cent Cu below which there is no longer a geometrical increase in copper reserves as mineable grade decreases.

(c) According to Figure 2, the answer must be no. Virtually all the low-grade copper deposits are the porphyry coppers, which suggests that exploration programmes will for some time yet be concentrated on presently or recently active destructive plate margins.

**ITQ 5** 1 (a) *Definitely has.* By providing a better understanding of crustal processes, plate tectonics has helped our understanding of ore genesis.

2 (b) *Probably has.* Plate tectonics has emphasized that examination of the global distribution of ore deposits can assist in mineral exploration and this is certainly important. However, it must also be remembered that the occurrence of porphyry copper deposits in orogenic belts and associated with granitic intrusive rocks was recognized before plate tectonics; plate tectonics so far has merely assisted our understanding of this environment in terms of plate motions.

3 (c) *Probably has not.* The search for porphyry copper deposits on a local scale is concerned with the 'nitty-gritty' of geology, such as faults and the nature of the rocks, rather than large-scale processes.

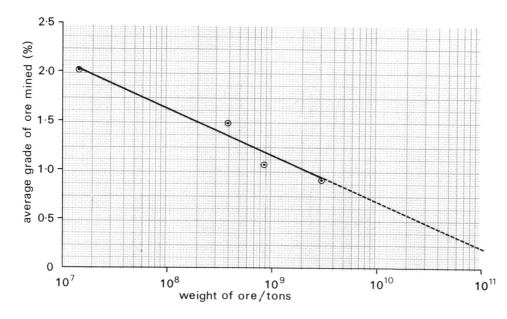

*Figure 50* Graph of mineable grade against tonnage of ore for the Bingham deposit, using data from Table 3.

**TABLE 16** Answer to ITQ 6.

| (a) Location of the ore | Recognition of surface guides to ore (surface colorations, old workings); reconnaissance mapping of structures, rock types, mineralization; construction of an accurate geological map; selection of drilling targets; exploratory drilling. |
|---|---|
| (b) Collection of geological data | Surface mapping using transit and plane table to plot topography; detailed recording of rock types, structures, and alteration facies; borehole logging; recording data from tunnels and channels; construction of underground maps of various levels and cross-sections. |
| (c) Estimation of the grade and extent of the orebody | Assay of rocks from boreholes, tunnels, channels and surface exposures; production of assay overlays to geological maps and sections; use of these overlays to make projections and predictions of the size and grade of the mineralized zone. |
| (d) Choice of mining method | Assessment of the size, shape, geometry and grade of the orebody; assessment of the physical properties (strength) of the rocks; comparison of the costs of the various possible methods. |

**ITQ 6**  See Table 16.

**ITQ 7**  (a) You should look for the presence of limonite and the general absence of sulphides as indicators of leached capping; turquoise colouring as an indicator of oxidation; the minerals covellite and chalcocite as indicators of supergene enrichment; and the absence of these features as indicators of primary ore.

| 0–50 m | leached capping and oxidation |
| 50–145 m | supergene enrichment |
| 145–190 m | primary ore |
| 190 m–end of core | further supergene enrichment |

(b)  You should have estimated:

| | Cu/per cent | Mo/per cent |
|---|---|---|
| Segment 1 | 0.1 | 0.01 |
| Segment 2 | 0.4 | 0.01 |
| Segment 3 | 0.2 | 0.01 |
| Segment 4 | 0.25 | 0.03 |

(i)  From these estimates you would be highly pleased if this were an exploratory hole.

(ii)  If this were in the centre of the deposit you would need to drill deeper to assess its true value; although the Cu concentration is probably only marginally economic, the high Mo concentrations at deep levels look promising.

**ITQ 8**  (a)  $K$-porphyry is more intensely mineralized.

(b)  The $L$-porphyry cross-cuts—and is, therefore, later than—the $K$-porphyry

(c)  The ore occurs in disseminated form, in small veinlets and in later, larger veins.

**ITQ 9**  In the case of (a) and (b), the rock should break up fairly easily, so less explosive should be needed. In the case of (c) and (d), the rocks will break up only with difficulty. Also, with clay alteration (b), some support may be needed to prevent collapse of underground workings.

**ITQ 10**  1 = B  3 = A  5 = C
2 = F  4 = E  6 = D

In Figure 51 we have summarized the geological interpretation of the geophysical data.

**ITQ 11**  (a) These porphyry copper deposits seem to be related to igneous rocks of Cretaceous–Tertiary age. This is also true for other such deposits in Chile, both north and south of the El Salvador region.

(b) The Jurassic rocks along the coast are deeply eroded and any porphyry copper deposits that they once contained may already have been eroded away.

(c) The Quaternary volcanic belt, by contrast, has not been eroded deeply enough. Any mineralized intrusive rocks lie beneath a thick cover of lavas.

(d) The Cu–U–V red-bed deposits, which you encountered in SAQ 1, are found in fluviatile sandstones of this type.

**ITQ 12**  (a) Upper Cretaceous to Lower Tertiary

(b) Tertiary

(c) Between Upper Cretaceous and Lower Tertiary

(d) Tertiary, before intrusion of the igneous rocks

(e) Tertiary, post volcanics

(f) Upper Cretaceous and Tertiary

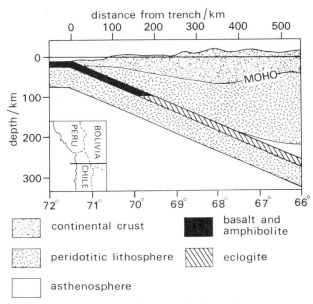

*Figure 51*  Geological section through the Andes.

**ITQ 13**  The answer is given in Figure 52.

**ITQ 14** (a) You should have produced an AFM diagram similar to that in Figure 53. Without having any analyses for basic rocks, it is difficult to compare the fractionation trend for El Salvador with that for Skye. There appears to be *no iron enrichment* and the trend, therefore, resembles the alkalic rather than the tholeiitic trend (*Handbook*, Section VI, Fig. 14).

(b) El Salvador shows (i) the greater volume of acid rocks, (ii) less $SiO_2$ and, in view of the higher MgO and FeO concentrations, (iii) a greater proportion of mafic minerals.